CW01316707

Backyard Meat Rabbits

A Comprehensive Guide to Raising Rabbits for Meat, Including Tips on Choosing a Breed, Building the Coop, and Harvesting

© Copyright 2023 - All rights reserved.

The content contained within this book may not be reproduced, duplicated, or transmitted without direct written permission from the author or the publisher.

Under no circumstances will any blame or legal responsibility be held against the publisher or author for any damages, reparation, or monetary loss due to the information contained within this book, either directly or indirectly.

Legal Notice:

This book is copyright-protected. It is only for personal use. You cannot amend, distribute, sell, use, quote, or paraphrase any part of the content within this book without the consent of the author or publisher.

Disclaimer Notice:

Please note the information contained within this document is for educational and entertainment purposes only. All effort has been executed to present accurate, up-to-date, reliable, and complete information. No warranties of any kind are declared or implied. Readers acknowledge that the author is not engaging in the rendering of legal, financial, medical, or professional advice. The content within this book has been derived from various sources. Please consult a licensed professional before attempting any techniques outlined in this book.

By reading this document, the reader agrees that under no circumstances is the author responsible for any losses, direct or indirect, that are incurred as a result of the use of the information contained within this document, including, but not limited to, errors, omissions, or inaccuracies.

Table of Contents

INTRODUCTION 1
CHAPTER 1: INTRODUCTION TO BACKYARD RABBIT FARMING 3
CHAPTER 2: CHOOSING THE RIGHT BREED FOR PRODUCTION 16
CHAPTER 3: CREATING A HEALTHY ENVIRONMENT 30
CHAPTER 4: UNDERSTANDING YOUR RABBITS' NUTRITIONAL NEEDS 39
CHAPTER 5: PREVENTING AND ADDRESSING HEALTH ISSUES 48
CHAPTER 6: ETHICAL BREEDING PRACTICES FOR SUSTAINABILITY 60
CHAPTER 7: THE LIFE CYCLE OF A RABBIT 69
CHAPTER 8: COMPASSIONATE HARVESTING 78
CHAPTER 9: USING RABBIT BY-PRODUCTS 86
BONUS CHAPTER: RESPONSIBLE RABBIT RAISING: ETHICS AND REGULATIONS 95
CONCLUSION 105
HERE'S ANOTHER BOOK BY DION ROSSER THAT YOU MIGHT LIKE 107
REFERENCES 108

Introduction

Why should you dive headfirst into the world of raising rabbits for meat? There are so many reasons to do so. It's been done forever, yet it's often overlooked. There was a time when rabbits weren't mere domestic companions but a vital part of the homestead's meat production. Long before backyard BBQs became synonymous with sizzling steaks and plump chicken drumsticks, people knew the hidden gem that was rabbits. Once wild and elusive, these fluffy creatures gradually found their place as house pets. It wasn't an overnight transformation, mind you. Before this happened, rabbits were mainly used as a food source.

Now, you may be wondering. Chickens and cows have long been the go-to options for meat, so why go for rabbits? Consider the efficiency of it all. Chickens and cows demand space, feed, and time—a trifecta of valuable resources. Rabbits, on the other hand, are compact powerhouses. They don't require sprawling pastures or massive feed bins. A small corner of your yard can become a rabbit haven and can yield an impressive harvest of tender meat.

And talk about speed. Chickens and cows take their sweet time to mature, demanding your patience while you wait for that perfect moment to savor their flavors. But rabbits? They're the sprinters of the meat world. In a matter of weeks, you'll have rabbit meat on your plate. It's a satisfyingly swift turnaround that even the most hurried homesteader can appreciate, but there's more. Imagine a life where your meat source is not only economical but also sustainable. Rabbits are known for their prodigious breeding abilities, and their swift reproduction cycle ensures a steady supply of meat for your table. While chickens and cows may require more attention to their reproductive needs, rabbits practically write their own script, creating a delightful surplus of meat.

Starting your rabbit adventure comes with various possibilities, and each path has its own goal. There is bound to be one that matches your interests and abilities. Maybe you're aiming for self-sufficiency. Think of having rabbits as a way to create a mini ecosystem. These little creatures can give you good protein to eat. Or perhaps you're into farmers' markets. Rabbit meat may not be the usual choice for some people, but once they taste it, they could become your loyal customers. Another option is connecting with restaurants. Rabbit has become quite popular on menus thanks to creative chefs. By supplying them with quality rabbit meat, you become part of this culinary trend.

However, *keep it real.* Be honest about what you can handle – your skills, time, and resources. Don't get swept up in excitement without a clear goal. There's a story about a family who bought rabbits to eat, but when it was time to process them, they couldn't. They ended up with pets instead of meat. One thing to remember from the start is that raising rabbits for meat means you'll eventually need to deal with harvesting them. It's a serious conversation to have upfront so you know if you're comfortable with the whole process before you even start mating rabbits.

In the world of raising rabbits for meat, your journey may have a different destination, but what ties it all together is your dedication to learning, providing, and maybe even processing rabbit meat. It's about finding your own path and enjoying the adventure along the way.

Chapter 1: Introduction to Backyard Rabbit Farming

Backyard rabbit farming has emerged as a gratifying pursuit for individuals and families looking for a harmonious blend of companionship and sustainable food production. With their docile nature, modest upkeep demands, and prolific breeding tendencies, rabbits have captured the interest of those looking to engage in small-scale animal husbandry. Farming rabbits is a perfect opportunity to cultivate a connection with these endearing creatures while enjoying the benefits of homegrown meat. As the book delves into backyard rabbit raising, it will navigate this venture's various considerations, techniques, and rewards, encompassing everything from selecting suitable rabbit breeds to creating sustainable habitats that fit your lifestyle.

Farming rabbits is a perfect opportunity to cultivate a connection with these endearing creatures while enjoying the benefits of homegrown meat.
https://pixabay.com/photos/rabbit-farmer-rabbit-pet-7657156/

Exploring Backyard Rabbit Farming

The realm of backyard rabbit farming is a multi-faceted journey that harmonizes nurturing living beings with sustainable agricultural practices. A fundamental aspect is selecting rabbit breeds that fit in with your goals: meat production, fur, or engaging pets. These decisions lay the foundation for a rewarding experience. It's equally crucial to provide rabbits with appropriate housing space. From cabinets to tractors, the well-being and protection of these animals are paramount considerations. Likewise, attending to their dietary needs with a balanced blend of fresh greens, hay, and nutritionally dense pellets ensures optimal health.

Beyond the practicalities, raising rabbits offers the enriching experience of observing natural behaviors, tending to the growth of rabbit families, and fostering a connection with the rhythms of nature. The benefits extend further, as your garden benefits from the valuable resource of rabbit-waste compost. Moreover, this undertaking encourages a deeper understanding of animal care, husbandry ethics, and sustainable living practices. As you embark on this exploration of backyard rabbit farming, you will find this book will uncover the nuances that make it a fulfilling and educational endeavor for those seeking a connection to the natural world and a pathway to sustainable, self-sufficient living.

Why Raise Backyard Rabbits for Meat

Rabbits have garnered attention as a practical and environmentally conscious option for those seeking an alternative to conventional meat production. Farming rabbits as a source of meat is based on factors like a short reproduction cycle, efficient conversion of feed to protein, and manageable space requirements. These factors have made rabbits a feasible pick for families and individuals aiming to embrace sustainable food practices while making a minimum ecological footprint.

Efficient Feed-to-Meat Conversion Rate

Rabbits stand out for their remarkable efficiency in converting feed into high-quality protein. Known for their herbivorous diet, they have a specialized digestive system that allows them to take the highest amount of nutrition from plant-based materials such as hay and grains. This efficient feed-to-meat conversion rate makes rabbit meat a lean and healthy option. It contributes to resource conservation by minimizing the amount of feed needed to produce a substantial amount of protein.

Minimal Space Requirements

Farming rabbits in your backyard is perfect for people with limited space. Unlike larger domestic animals that demand extensive grazing areas, rabbits can thrive in modest enclosures such as hutches or pens. This adaptability to confined spaces is particularly appealing in urban and suburban settings where land is scarce. Consequently, raising rabbits provides an avenue for meat production even in environments where traditional livestock farming would be impractical.

Rabbits can thrive in modest enclosures such as hutches or pens.
https://pixabay.com/photos/rabbit-hutch-house-easter-cottage-502929/

Dual-Purpose Nature: Meat and Fur

Another compelling facet of rabbit raising is the dual-purpose nature of meat and fur production. Besides providing tender and flavorful meat, certain rabbit breeds' soft and dense fur can be used to make knitted goods like gloves, cardigans, and felts. However, if the fur length is too short, this makes it impossible to make yarn. This dual functionality aligns with sustainable practices by maximizing the yield from each animal, reducing waste, and supporting local artisanal endeavors, choosing rabbits suited to both meat and wool production.

Rapid Reproduction Cycle

The swift reproductive cycle of rabbits contributes to their appeal as a source of meat. A single doe (female rabbit) can produce multiple litters of kits (baby rabbits) each year, resulting in a consistent meat supply. This reproductive efficiency allows for a sustainable and predictable meat production rhythm, reducing the time and resources required to yield a substantial harvest.

Lower Environmental Impact

Rabbit farming dovetails with environmentally conscious practices due to its reduced ecological footprint. Compared to larger livestock, rabbits consume less feed, require smaller living spaces, and generate fewer greenhouse gas emissions. Their efficient resource consumption contributes to conservation efforts by minimizing water usage and land requirements.

Health Benefits

Rabbit meat is considered a healthy protein option because it is a low-fat and low-cholesterol protein. This makes it an appealing choice for people aiming to maintain their weight and cardiovascular health. The lean rabbit meat has a high protein content that encourages muscle health and improves immune function and overall well-being.

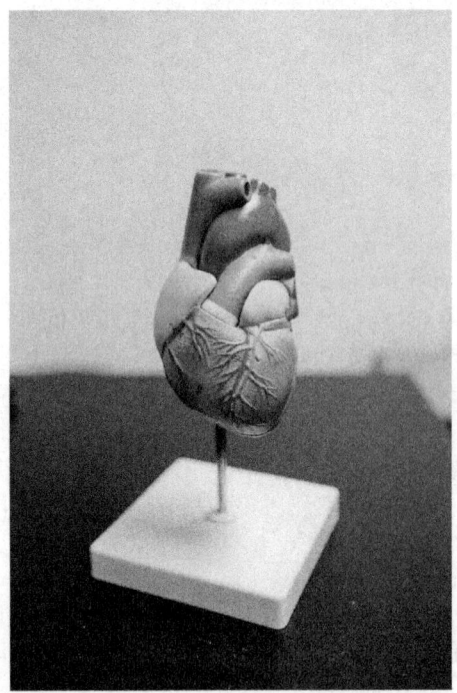

Rabbit meat is an optimal choice for people aiming to maintain their weight and cardiovascular health.
https://unsplash.com/photos/MEbT27ZrtdE

Nutrient-Rich Meat

The meat from rabbits is abundant in essential nutrients that play crucial roles in maintaining metabolic processes. For example, the meat is packed with B vitamins, particularly B12, which is a vital substance required for energy metabolism and nerve function. Likewise, the high iron, zinc, and phosphorus in the meat improves the transport of oxygen in the blood, zinc supports immune system health, and phosphorus improves bone health and cellular function.

Economic Viability

Raising rabbits can be economically practical, making it an accessible option for those looking to produce their own meat. Rabbits grow quickly and efficiently convert feed into meat, resulting in a relatively high protein yield from a modest investment. This efficiency contributes to cost-effective meat production.

Accessibility to Urban Dwellers

The adaptability of rabbits to confined spaces makes them a viable choice for urban and suburban areas with limited land availability. Unlike larger livestock, rabbits can thrive in smaller enclosures. This accessibility enables urban dwellers to produce meat without having to own vast tracts of land.

Educational Value

Rabbit farming also offers an educational opportunity, particularly for children. Caring for rabbits teaches responsibility, empathy, and practical skills. Children can learn about animal care, life cycles, biology, and the importance of treating animals with kindness and respect.

Ethical Considerations

For individuals who prioritize the ethical treatment of animals, raising rabbits aligns with their values. The manageable size of rabbits makes them less intimidating to handle than larger livestock. This can lead to a more humane and less stressful experience when raising and harvesting rabbits for meat and fur.

Reduced Antibiotic Use

Raising rabbits in your backyard involves fewer antibiotics than large-scale commercial meat production. Rabbits are generally hardy animals with fewer health issues, and because of their small size, they get more individualized care, reducing the need for routine antibiotic use.

Local Food Security

Rabbit farming contributes to local food security, ensuring a consistent supply of fresh meat within communities. This localized production decreases dependence on distant food sources during local market supply disruptions, as most rabbit species thrive well in extreme temperatures. This ability to survive in hot and cold temperatures makes rabbits a tremendous alternative meat source. Furthermore, rabbits can adapt to different terrains, making it easier to raise them wherever humans live.

Customized Breeding Programs

Raising rabbits allows breeders to tailor their breeding programs to meet specific goals, whether it's to optimize meat yield and fur quality or adapt to local climates. This customization provides opportunities for experimentation and innovation. Successful breeding programs can further lead to the development of species with better qualities in any of these areas.

Homestead Diversification

For those pursuing a self-sufficient lifestyle, rabbits can be a valuable addition to a diversified homestead. Integrating rabbit husbandry with other practices such as gardening, poultry, and small livestock contributes to a greater variety of resources available for personal consumption.

Hands-on Sustainability

Being such a hands-on and simplified farming activity means that a greater connection is made with the food source, encouraging sustainable practices. Individuals become more mindful of their food consumption and the resources required to produce it, promoting a deeper understanding of sustainability.

Connection to Natural Cycles

When you engage in raising rabbits, this offers insights into the natural cycles of life, reproduction, and responsible animal stewardship. This experience enhances relevant knowledge and awareness of the processes that sustain life and a deeper appreciation for the natural world.

Preservation of Heritage Breeds

Yet another positive for backyard farming is the all-important preservation of genetic diversity and cultural heritage of heritage rabbit breeds. This support for biodiversity safeguards unique breeds from extinction and maintains their historical significance.

Empowerment and Resilience

Rabbit raising empowers small farmers to take control of their food sources and become more self-reliant. Small farmers can produce food and be less dependent on external supply chains while raising rabbits. This process contributes exponentially to personal empowerment.

The decision to consider rabbits as a source of meat embodies a multifaceted approach to sustainable food production. Their efficient feed conversion, minimal space needs, dual-purpose nature, rapid reproduction cycle, and lower environmental impact converge to offer a practical and ethical choice for people seeking to nourish themselves while prioritizing conservation and self-sufficiency responsibly. As an alternative to conventional meat sources, rabbits exemplify how mindful decisions in food production can contribute to a more sustainable and harmonious relationship with the environment.

Addressing Challenges in Backyard Rabbit Farming

Starting your journey on the path to farming rabbit meat has some challenges that require careful attention and proactive management to ensure the well-being of the rabbits and the success of your venture. Here's a detailed exploration of these challenges and how to address them:

Starting your journey on the path to farming rabbit meat has some challenges that require careful attention.
https://unsplash.com/photos/bJhT_8nbUA0

1. Appropriate Housing

Rabbits in their natural habitat are fond of digging underground warrens (tunnels). When raising rabbits in an urban or residential space, it's crucial to provide suitable housing for the rabbits' safety and comfort. Hutches or pens should shield rabbits from predators, provide shelter from weather extremes, and offer proper ventilation. Insulating the housing helps regulate temperature, and correct spacing between wires prevents injuries. Regular cleaning is essential to avoid waste buildup, which leads to health issues and unpleasant odors.

2. Health and Veterinary Care

Maintaining rabbit health requires regular monitoring. Common health concerns in rabbits include digestion issues, dental issues, respiratory infections, and external parasite attacks. Besides conducting regular checkups by yourself, it's better to call in a certified veterinarian who can examine rabbits for signs of illness and behavioral changes and note down changes in diet and stool quality. They are very well-trained to recognize symptoms and provide immediate medical attention to achieve the best results.

3. Meeting Dietary Requirements

Meeting rabbits' dietary needs is vital for their well-being. Their diet should consist of high-quality hay like timothy or orchard grass, fresh vegetables such as leafy greens and carrots, and commercially balanced rabbit pellets. Avoid feeding rabbits foods high in sugar or low in fiber, as these foods can increase toxicity. Ensure access to clean, fresh water at all times to prevent dehydration.

4. Reproduction Management

While rabbits' breeding capacity is advantageous, it must be managed carefully. Uncontrolled breeding can lead to overpopulation, stress, and compromised well-being for rabbits and caretakers. Put into place a breeding plan that you can manage and control. Separate males and females to prevent unintentional breeding.

5. Social and Behavioral Needs

Rabbits are social animals that thrive on companionship. However, introducing rabbits requires a gradual and monitored process to prevent aggression. Housing rabbits alone can lead to loneliness and behavioral problems. Introduce rabbits on neutral territory, monitor interactions, and initially keep them separated in order to avoid conflict.

6. Predator Protection

Rabbits are natural prey animals, making them vulnerable to predators. Secure enclosures with sturdy fencing depending on the predator threat, appropriate wire spacing in fences, and solid barriers help deter predators. Consider adding further predator deterrents like motion-activated lights or noise-making devices.

7. Environmental Enrichment

Rabbits are intelligent and curious creatures that need mental and physical stimulation.

Lack of enrichment can lead to boredom and unwanted behaviors. If you are raising rabbits in an urban space, provide toys like cardboard boxes, create tunnels, and chew toys to keep them occupied. Offer hiding spots, platforms, and opportunities for digging to mimic their natural behaviors.

8. Waste Management

Adequate waste management is crucial to maintaining a healthy living environment. Regularly cleaning soiled areas, cages, and hutches and disposing of waste responsibly are some basic steps for waste management. Incorporating waste management techniques effectively prevents the development of foul odors, reduces the risk of disease transmission, and discourages fly or other pest infestations.

9. Climate Considerations

Extreme temperatures can impact rabbit health. Ensure your housing has adequate ventilation and insulation to prevent heat stress or cold-related health issues. Offer shade in hot weather and warmth in cold weather. Monitoring weather forecasts and making necessary adjustments to their living environment is crucial.

Be sure to offer your rabbits shade in hot weather and warmth in cold weather.
https://www.pexels.com/photo/rural-snowy-village-during-severe-blizzard-4969828/

10. Learning and Adaptation

Raising rabbits is a learning curve, especially for people unfamiliar with animal husbandry. Educating yourself about rabbit care, behavior, and needs through books, online resources, and advice from experienced rabbit owners is best. Be open to adapting your practices based on what works best for your rabbits, as each rabbit species may have unique preferences and requirements. Likewise, the space you keep them in also defines their individual requirements.

11. Parasite Prevention

Regularly inspect your rabbits for signs of external parasite infestation. You'll notice rabbits scratching their fur due to itching, fur loss, or visible pests on their fur. To prevent parasites, always keep their living space ventilated, dry, and clean. Regularly change bedding, clean enclosures, and provide fresh hay. When unsure, don't hesitate to consult a veterinarian for appropriate preventive measures or treatments if necessary.

12. Handling and Socialization

Gentle and positive handling is crucial for rabbits' well-being. When picking up a rabbit, support its hindquarters to avoid injury. Spend time near them, offering treats and gentle strokes. Gradually increase interaction to help them become accustomed to your presence and build trust.

13. Quarantine Procedures

When introducing new rabbits to your existing group, implement a quarantine period of about two to four weeks. This minimizes the risk of introducing diseases. Keep new rabbits separate during this time and monitor their health closely. Consult a veterinarian for guidance on quarantine procedures.

14. Observing Behavior

Regularly check your rabbits' behavior to detect any changes. Rabbits are experts at hiding signs of illness, so any alterations in eating habits, activity levels, grooming, or behavior could indicate underlying health issues. Promptly address any concerning changes.

15. Safety from Chemicals

Rabbits like to nibble on things, so ensure their environment is free from toxic substances and plants. Remove any chemicals, pesticides, or potentially harmful materials from their living area to prevent accidental ingestion.

16. Handling Stress

Sudden changes or disturbances can trigger stress in rabbits. You can minimize stress in these sensitive creatures by providing a stable environment, avoiding loud noises and sudden movements, handling them gently, and limiting prolonged exposure to unfamiliar sounds.

17. Grooming Needs

Long-fur rabbits like Angoras need regular grooming to prevent fur matting and tangling. Not fulfilling their grooming needs will cause discomfort and lead to several other skin issues. It's crucial to understand that the grooming requirements change slightly with every rabbit species. Recognizing and fulfilling these grooming needs is your responsibility as their caretaker. Rabbits with long and thick fur will need more care and attention than short-haired breeds.

18. Fostering Trust

Rabbits are intelligent and can bond with humans caring for them, but it takes time and patience. To foster trust, try spending more time near their enclosure and feed them their favorite snacks once a day. You can offer treats or fresh vegetables by hand to create positive associations. Avoid forcing interaction and allow them to approach you at their own pace.

19. Health Records

Keep accurate health records for each rabbit. Document their medical history, previous treatments, vaccination data, and any health concerns they are suffering from. These records are valuable for tracking their health, discussing concerns with veterinarians, and making informed breeding and care decisions.

20. Community and Resources

Talking with other rabbit enthusiasts through joining local or online communities will benefit you exponentially. You can share your experiences, seek advice from veterans in rabbit farming, and educate yourself on other's expertise. Interacting with experienced rabbit owners will undoubtedly provide valuable insights and support.

21. Time Commitment

Farming rabbits takes time, dedication, and commitment. Your daily tasks will include feeding, cleaning enclosures, monitoring health, and providing social interaction. Be prepared to dedicate time to their care, as neglecting their needs can lead to health issues and poor well-being.

22. Emergency Preparedness

Always be prepared for any medical emergency with a feasible action plan. The plan can include knowing evacuation procedures, providing first aid, and having contact information for a veterinarian knowledgeable about rabbit care. Being prepared ensures a swift response in critical situations.

23. End-of-Life Considerations

Understanding the rabbits' end-of-life needs and acting humanely and responsibly is crucial in rabbit farming. If a rabbit is suffering from a terminal illness or has their health deteriorating rapidly, be prepared to make difficult decisions about euthanasia in consultation with a veterinarian. Have a plan for proper disposal and consider environmentally friendly methods if needed.

By addressing these points, you'll be well-prepared to navigate the challenges of backyard rabbit farming. Taking a proactive and informed approach ensures the well-being of the rabbits and promotes a positive and fulfilling experience for both the caretaker and the rabbits themselves.

Imagine a sustainable food production method that fits into your backyard, offers nutrient-rich meat, and introduces you to a world of unique companionship. Backyard rabbit farming for meat isn't just a venture; it's a journey that connects you to the rhythms of nature, nourishes your curiosity, and enriches your understanding of ethical food sources.

As you step into raising rabbits, you're entering a world where efficiency meets compassion. Discover how these small, cute, and furry creatures have made waves in the culinary industry, having abilities to convert feed into high-quality protein with nutrient-rich deposits. Explore the intricate dance between sustainable practices and responsible stewardship as you embark on a path that transcends traditional meat production methods.

Picture yourself creating tailored living spaces that provide comfort and security for your rabbits, and witness the joy of nurturing lives that, in turn, nourish you. As you explore the details of housing design, health maintenance, and dietary needs, you'll uncover the fascinating intricacies of rabbit care. Each challenge you face becomes an opportunity to deepen your connection with these creatures and enhance their quality of life.

Imagine the satisfaction of taking control of your food source, knowing that the meat on your table has been raised with care and integrity. Backyard rabbit raising isn't just about sustenance; it's a holistic approach that touches upon health benefits, ethical considerations, and the satisfaction of being part of a community of responsible food producers.

The world of rabbit farming invites you to explore beyond the confines of traditional meat consumption. It encourages you to embrace a hands-on, sustainable lifestyle that aligns with nature's rhythms. Whether you're a novice or a seasoned enthusiast, this journey is about fostering a deep connection with the animals you raise, the environment you nurture, and the sustenance you derive from it all.

Are you intrigued to learn more about the art and science of breeding rabbits for meat? Dive into the captivating world of responsible food production, compassionate care, and sustainable living. Discover how the humble rabbit can be a source of culinary delight and a profound connection to the natural world. Your journey into backyard rabbit raising promises a tapestry of experiences that enrich your life while contributing to a healthier planet.

Chapter 2: Choosing the Right Breed for Production

Knowing the right rabbit breed to choose from for your meat production is essential. If you are planning on going into it commercially, it can lead to a lucrative and rewarding business venture. Since no two breeds of rabbits are the same based on their characteristics difference, it implies that their adaptability, litter size, growth rate, feed conversion, and the quality of meat they produce would also differ. Therefore, knowing and going after the rabbit breed appropriate for your farming needs and goals is essential to successful production.

This chapter provides you with the different rabbit breeds suitable for meat production. You will discover the factors that make these rabbit breeds suitable, along with some popular meat breeds and their unique needs.

Various Rabbit Breeds Most Suitable for Meat Production

New Zealand White

New Zealand White is a known meat-producing breed.
https://commons.wikimedia.org/wiki/File:NewZealandWhiteRabbit_2.jpg

New Zealand White is a known meat-producing breed. It grows very fast, and its meat is tasty and soft. The amount of meat in a New Zealand White rabbit is more than its bone, and the meat flavor is excellent. The New Zealand White has a large litter size, that is, eight to twelve kits per litter on average, coupled with an exceptional feed conversion ratio and a rapid growth rate. These traits make the New Zealand White an effective breed suitable for meat production.

For adaptability, the New Zealand White can be reared in different climates, and they are very easy to look after. Additionally, you can easily manage them because they are compliant, which makes them a good option for anyone new to rabbit farming.

The New Zealand breed is certainly the breed for meat production due to its fast growth rate, yield, and overall management.

Californian

Another well-known breed for meat production, with a weight growth of eight to 12 pounds in 12 weeks, is the Californian. This breed is known for its tender and flavorful meat. It has an exceptional meat-to-bone ratio, which is sought after by restaurants and butchers.

Furthermore, the Californian grows rapidly, and its good conversion rate makes it an effective breed for producing meat. The average litter size of the Californian is six to eight per litter, and their growth rate is similar to that of the New Zealand White.

As for the ability to adapt, you can easily rear the Californian breed in any climate, and they are easy to nurture. The Californian breed has a gentle disposition. It is a crossbreed between the New Zealand Whites and the Chinchilla rabbits.

American Chinchilla

Due to the popularity of their meat and fur, this breed is called a dual-purpose rabbit. With a weight of over 12 pounds and a stocky body, they are regarded as one of the finest meat breed rabbits in the world. People favor this breed for its broad shoulder and superior deep loin, seen in various smoked and cooked dishes worldwide. Due to its popularity, this breed of rabbit is considered endangered.

American Chinchillas make good mothers and are known to give birth to eight to twelve kits. Additionally, they are very friendly, weigh between nine and 12 pounds, and have an exceptional meat-to-bone ratio.

Rex

Rex is considered a popular breed for meat production.
DestinationFearFan, CC BY-SA 4.0 <https://creativecommons.org/licenses/by-sa/4.0>, via Wikimedia Commons: https://commons.wikimedia.org/wiki/File:Rex_rabbit_(calico).jpg

Rex is also considered a popular breed for meat production. It weighs seven to 11 pounds and averages six to 12 kits per litter. This breed can be purchased easily in the US and is appreciated for its velvety pelt and meat-to-bone ratio. However, the Rex breed takes longer to get to the dinner table when compared with the New Zealand white.

Champagne D'Argent

The Champagne D'Argent breed is well-regarded around the world. This breed is known as the godfather of rabbits and has been a source of meat since 1631. Champagne D'Argent got its name from the city of Champagne in France, where it originated. A mature Champagne D'Argent has a weight of nine pounds with an uncommon amount of bone-to-meat. You can get both meat and fur from a Champagne D'Argent breed.

Silver Fox

The Silver Fox is known for its good meat quality and fur among small-scale farmers. Within three months, they can reach 10 to 12 pounds. They usually have medium-sized litters of between seven and eight kits. They are also rare, as they are considered an endangered breed. Individuals skilled in tanning hides highly prize the silver Fox breed's stunning pelt.

Satin Rabbits

The Satin breed is regarded as one of the heaviest and largest rabbit breeds, weighing over 12 pounds when fully grown. This rabbit produces a reasonable amount of meat due to its larger body size. Satins have a docile and calm temperament. They are the ideal breed of meat rabbits to nurture on your farm.

Cinnamon Rabbits

This breed is a cross between the New Zealand and American chinchilla rabbit. Although the initial purpose of this rabbit wasn't to produce meat, with its 11-pound weight when fully mature, it is certainly a breed to be considered for commercial purposes. The Cinnamon rabbit is red, prized for its fur, and can be kept as a pet. Nevertheless, this breed is hard to find.

Palomino Rabbit

Palomino rabbits have been known as meat rabbits for decades.
Jamaltby at en.Wikipedia, CC BY-SA 3.0 <https://creativecommons.org/licenses/by-sa/3.0>, via Wikimedia Commons: https://commons.wikimedia.org/wiki/File:PalBuckSide-small.jpg

This breed has been known as a meat rabbit for decades, popular for producing meat for both subsistence and commercial purposes. Palomino rabbits weigh 8 to 11 pounds when mature and have an excellent meat-to-bone ratio. The palomino rabbit has an easy-going temperament, which is why you can raise them. Nevertheless, you must be patient with them as their growth process is usually slow compared to other meat-producing rabbits.

American Blue

The American Blue rabbit can give you both meat and fur. It weighs nine to 12 pounds, and being an excellent mother, it averages 8 to 10 kits per litter. Sadly, it has a poor meat-to-bone rate like the Flemish Giant and is better off when crossbred with other breeds such as Silver Fox, Harlequin, Rex, or any other smaller breed.

Factors That Make a Rabbit Breed Suitable

There are certain factors to check for when choosing a suitable breed for meat production. These include growth rate, typical adult sizes, and overall temperaments.

Growth Rate

The growth rate of a breed is vital when considering a rabbit breed for your farm. This is key because a rapidly growing rabbit produces an early harvest, leading to regular meat production. When dealing with growth rates, consider quality breeding stock. Choose rabbits from thrifty lines that can quickly pay for themselves in feed savings alone. Seek out breeders with a good track record and purchase from them. Remember to look elsewhere if the breeder you are dealing with cannot tell you what their kits weigh at eight weeks.

Additionally, when choosing rabbits with a good growth rate, find rabbits with quality meat bloodlines. Good meat bloodlines are breeds of rabbits selected over generations for their meaty body type, rapid growth rate, and thriftiness. The offspring of these fast growers also tend to mimic these great qualities.

Furthermore, go for commercial rabbit breeds that have a fine bone structure. The offspring of these breeds are known to be speedy growers with an excellent meat-to-bone ratio, mainly from a 60 to 65 % dress-out rate. At maturity, adults weigh eight to twelve pounds.

Don't make the mistake of adding large-boned rabbits like Flemish Giants to your meat breeding program if you're looking for a faster growth rate. If you do, you will likely have a 5# fryer at eight weeks, which may not have much meat due to its large structure. Since rabbits grow bones before they grow meat, they may clean you out at five to six months old before they develop enough meat to account for their butchering. New Zealand and California breeds are well-known for commercial production because of their fast growth rate and litter size. On the other hand, heritage breeds are suitable for meat production in backyards and small farms.

Typical Adult Sizes

Most people commonly believe rabbits ought to be small pets. Due to this assumption, they are surprised when the baby rabbit they bring home turns into a giant rabbit the size of a cat! For example, the Holland Lop is a small breed of domestic rabbit, but many people find it big. Even "dwarf" and "mini" rabbit breed sizes may be considered big to some people, as they can become up to five pounds.

The Average Size of Adult Rabbits

Rabbits vary in size based on their breed and age. Therefore, while the average-sized adult house rabbit weighs six pounds, it doesn't help you to imagine the size your backyard meat rabbit will grow to. Some rabbits are large, while others are small; get the facts before you choose your breed.

- **Small Rabbits**

Small rabbits comprise mostly the mini and dwarf breed rabbits. The weight of these rabbits will never exceed five pounds. Surprisingly, this category has the fewest number of rabbit breeds. The American Rabbit Breeders Association (ARBA) recognizes 50 rabbit breeds, out of which only 11 are under the 5-pound category weight bracket.

These rabbits are considered house pets because they are bred for their small sizes.

Small rabbits comprise mostly the mini and dwarf breed rabbits.
https://pixabay.com/photos/rabbit-bunny-easter-grass-cute-4813172/

- **Medium Rabbits**

Most of the commonly known rabbits are in the medium category. The adult weight of these rabbits is around five to eight pounds. You will see most rabbits weigh around five or six pounds, which is lower when compared to the given range. Even though medium rabbits are smaller than other rabbit breeds, they have an average size of two to three times bigger than most people anticipate for a backyard rabbit. Fifteen breeds of rabbits fall into this category.

- **Large Rabbits**

Rabbits in the large category have a typical adult size of eight to 15 pounds! The majority of these larger breeds are fostered mainly as meat-producing rabbits. However, even though most ARBA-recognized breeds are large rabbits, they are difficult to come by.

How to Know the Typical Adult Size of Your Rabbit

If you have a baby rabbit and are eager to find out how big it will get, there are simple ways to estimate it. Knowing the estimated size will enable you to prepare enough space for when they reach their full adult size.

- **Consider Their Breed**

An effective way to estimate the typical adult size of your rabbit is to consider their breed. A rabbit breed chart online will guide you on the size range to expect from your rabbit if you already know their breed.

- **Consider Their Age**

If your rabbit was adopted and it is hard to tell the breed, you can still estimate their typical adult size based on their age. Knowing your rabbit's current age will be a pointer for their weight. These tips will give you a close estimate of what to expect from the expected adult size of your rabbit.

- When your rabbit is about four months old, it will probably be half its adult size. For instance, if your small rabbit is currently three pounds, it will grow to be around six pounds as an adult.
- Your rabbit will probably be ⅔ of its adult size when it is over six to eight months old. For example, your adopted rabbit is not a year old and certainly not a baby; they will grow up a little. If your rabbit is three pounds at this age, their adult size will most likely be around 4.5 pounds.

Overall Temperaments

There are many misconceptions about rabbits. The most common one is that rabbits like to be held and cuddled because of their plush toy look. On the contrary, rabbits become active and assert their personalities once they reach sexual maturity. When they reach that stage, some people get rid of them because of limited information on how to raise them.

Rabbits have wide-ranging personalities, even among their littermates. They can be high-spirited, shy, timid, curious, gentle, and silly, regardless of breed type or sex. They show affection by climbing on your back, nibbling at your socks, or sitting close to you. Some can even go as far as licking your face or hand. Even belligerent rabbits can become affectionate toward you when given room to bloom.

The act of neutering can eliminate many behavioral problems and diseases in rabbits. Compared to larger rabbits, smaller and dwarf rabbits are more active than their older counterparts. Due to their lightweight, they can jump higher than the larger ones. A neutered backyard rabbit has an eight to 10-year average lifespan, though it tends to exceed that.

During their adolescent stage, rabbits display behaviors such as biting, spraying, nest building, loss of house training, nipping, courtship behaviors, and destructiveness like circling and mounting. Exhibiting these behaviors is not a sign of something wrong with your rabbit; it is typical development behavior. Consult a specialized veterinarian to get them neutered.

Furthermore, biting is your rabbit's way of relaying messages like bossiness, irritation, fear, lust, and curiosity. With a nip, rabbits tell each other to get out of the way! Do not offer your hand to your rabbit as a greeting or playful gesture. Your rabbit could interpret that as an intrusion or a threat.

Popular Meat Breeds, Their Special Needs, and Considerations

- **New Zealand White**

Physical Characteristics

New Zealand white is the most popular among the diverse varieties of the New Zealand breeds. It has a pure white color with bright pink eyes. New Zealand Whites have round cheeks, muscular faces, and slender, well-rounded bodies. They have small, short pectoral muscles and large, long back feet. Its average body weight is up to 11 lb.

Housing

New Zealand whites are best kept indoors to protect them from extreme weather and predators. Do not house your rabbits in areas you rarely frequent, as rabbits are social animals and enjoy company. Use a pen that is four times the length of your rabbit when it stretches. To give your rabbit a larger space, use dog playpens, which are bigger compared to commercial rabbit cages that are insufficient to house your rabbits. For at least five hours a day, give your rabbits the liberty to leave their cage. It will enable them to remain in a room without a pen or wander freely in your house.

See to it that the space allotted to your rabbits is rabbit-proofed! New Zealand whites have a natural tendency to chew and dig their way out, which can damage property like curtains, cords, rugs, furniture, and carpet. Also, make sure electrical wires are out of their reach.

Feeding

New Zealand Whites need a lot of fresh hay and water. Use hays like Timothy hay or other mixed grass hays. Every day, ensure you give your rabbit fresh, leafy vegetables. Supply at least one-fourth cup of vegetables for every pound of body weight. Some of the vegetables that a New Zealand White rabbit needs include carrot tops, bok choy red lettuce, and dandelion green.

Furthermore, you can use herbs like parsley, mint, cilantro, and basil to feed your rabbit. Non-leafy vegetables are considered unsafe for your rabbit, so abstain from feeding them. Feed your New Zealand White rabbits less carrots because of their high sugar level. Consult your veterinarian if you are unsure what food to give your rabbits.

Furthermore, you can use pellets from the store to complement your New Zealand Whites' diet. A timothy pellet with an 18% fiber content would suffice for an adult rabbit. However, the pellets should form a small amount of their diet, and you should never exceed the amount shown on the pellet's package.

Small amounts of fruit like pears, berries, melon, and apples can be given to your rabbit—one tablespoon for every three pounds of body weight.

Breeding

Breeding New Zealand Whites is simple. A doe becomes fertile between 8 and 12 weeks of age and can be bred at five to eight months. They are fertile throughout the year and have a gestation period of 28 to 35 days. However, most births take place at 31 to 32 days.

Care

Nurturing your rabbits will keep them healthy and encourage them to be productive.

Uses

The New Zealand White's primary use is meat production. They are fast growers, with their offspring (fryers) slaughtered at two months old.

Their fur is used to make fur trimmings in the fashion industry. Apart from their meat and fur production, the New Zealand White is used for commercial rabbit farming and for raising as pets.

Personality

New Zealand White breeds are outgoing and calm. They live well with both humans and other rabbits, making them social. They are a good option as pets because they can be handled easier than smaller breeds. Ensure you watch when children and other pets are close to your rabbits. Sudden movements and loud noises stress them easily. When they are not neutered, they act in a territorial manner.

- **Californian Rabbits**

Californian rabbits are reared for either their fur or their meat and are suitable as pets.
https://commons.wikimedia.org/wiki/File:Californian_Rabbit.JPG

Californian rabbits are among the most commercially bred rabbits in the US. They are often reared for either their fur or their meat and are suitable as pets.

Physical Characteristics

The Californian breed has a well-rounded, compact body and is large.

The Californian rabbit is similar in color to the Himalayan, having colored points with a white body. Its ears are big and stand erect. A brown marking is found on their tail, feet, ears, and nose. Californian rabbits have pink eyes and very short necks. They have full shoulders and are very muscular. They also have a silky and soft coat.

The Californian adult weighs 12 lb. on average.

Housing

Californian rabbits can either live in the house or outside the house. When going for a cage, ensure it's wide and long, with enough space for hopping and jumping. It will keep them healthier and happier.

It's normal for rabbits to chew their cage. This is why it is wise to ensure that the cage materials cannot easily be broken by your rabbit. A cage with a metal frame, a plastic bottom, or a metal bar with wire enclosing the sides can be cleaned easily and will keep the rabbit from destroying your home. If your cage is an all-wire design, ensure there is a resting place for your rabbit to avoid hurting its feet when sitting on the wire base. A nesting area built into the cage would do the trick, as it would help the rabbits avoid contact with the cage's base.

Feeding

Californian rabbits require a diet of vitamins A, D, and E and fiber. They need enough fat and protein. Pregnant or nursing females and growing rabbits need more protein than mature rabbits.

You can feed your rabbits fresh water, plenty of timothy hay, and a little pellet feed to give them the required nutrients. A matured Californian rabbit needs half a cup of pellets daily.

Feed your rabbits a lot of leafy vegetables with small amounts of carbohydrates. Some food items you can use include Radicchio, apples, pear, kale, green peppers, berries, broccoli, bok choy, etc.

Refrain from feeding your Californian rabbits food high in calories, seeds or nuts, grains, cookies, and bread. If you must give your rabbit carrots, do so in small amounts, as too many carrots can cause them harm.

Breeding

Californian rabbits can breed without human support. This rabbit has a gestation period of 28 to 31 days. A doe can give birth to two to eight kits at a time.

Care

Taking care of your rabbit is vital, especially if you are into commercial rabbit farming. Ensure that the pregnant nursing mother and breeding male are healthy. Regularly check them and always seek the services of a good vet while caring for them.

Uses

Californian rabbits are used for both their fur value and for meat production. It's an excellent breed to go for when going into commercial rabbit production. Furthermore, you can either raise them as show rabbits or pets.

Personality

Californian rabbits have a docile personality and can easily be managed.

- ## Flemish Giant

The Flemish Giant is the world's largest breed and one of the oldest. This breed is adaptable, gentle, and reared for its meat and fur.

Physical Characteristics

Apart from being one of the largest breeds in the world, their bodies are long with a broad back and solid, fleshed, well-rounded hindquarters. Although strong and muscular, their legs are of average length. Their ears are large and positioned in a V shape above their head. The head of the male Flemish Giant is broader and more imposing than that of the female.

Flemish Giants have a dense undercoat and smooth, medium-length hair with a glossy sheen. They exist in many varieties around the world. According to ARBA, there are seven color varieties: White, Sandy, Light Gray, Steel Gray, Fawn, Black, and Blue.

Flemish Giants have a body weight of up to 22 lb. However, the minimum standard weight for a senior buck is around 13 lb., while a senior doe is about 14 lb.

Housing

Due to their size, design a bigger cage with a larger enclosure. The minimum size for their cage is 3 x 4 feet. Smaller cages would cause them to be stressed because of their size.

Feeding

Since they don't overeat, feeding them with special commercialized pellets should do. Additionally, you can feed them cabbage, carrots, potatoes, parsley, pineapple, strawberries, corn, etc. Introduce these foods one at a time until their digestive system is accustomed to them.

For every five pounds of weight, Flemish Giants should be given two to four cups of vegetables daily. Put fresh water in their cage daily.

Breeding

Eight-month-old female Flemish giants are old enough to give birth after the 31st day. Its litter size averages 5 to 12 per litter. Since they are an old breed of domesticated rabbit, Flemish giants find it hard to breed with wild rabbits because of differences in their respective chromosomes.

Care

Flemish giants are gentle, obedient, and can adapt to any home. Although they can be considered pets, caution should be taken when kids are around them because they bite when they feel threatened or upset.

Uses

The Flemish Giant rabbit is well-known as a pet. Furthermore, they are suitable for producing meat and fur. This breed is also a popular show animal.

Personality

Though their looks and large size can throw you off, they are gentle giants. They love attention and are very friendly. They are peaceful creatures and crave a peaceful life. This doesn't mean you can push them around. They may scratch or bite you if you mistreat them.

Though daunting, choosing the right breed of rabbits for meat production is a wonderful experience. You now know the right breeds for meat production and how to get the best out of them. Enjoy your journey into raising the world's next source of meat.

Chapter 3: Creating a Healthy Environment

Rabbits are fragile yet active animals. They need to be kept healthy and comfortable. If you're new to rabbit farming, then there are terms you'll need to know. First, rabbit farming can also be called "cuniculture." Cuniculture allows you to breed domestic rabbits as a source of meat, fur, or both. To successfully raise these livestock, you must create a healthy and safe space where they can feed, breathe, nest, and breed. Rearing rabbits is an uncomplicated business that requires very few materials and resources. Rabbits eat almost anything nutritious. So, if you're excited and enthusiastic about starting this fruitful journey, this chapter provides the steps to follow.

You'll discover the secret behind raising your rabbits and doing it brilliantly. You'll also discover the space needed for a start-up, the suitable materials, and the best location for breeding, and learn how to create and maintain a hygienic environment. Apart from these elements, other factors, such as protecting your rabbits from predators and diseases, will also be discussed.

What Do Rabbits Need for a Comfortable Home?

Rabbits need enough room for a hop, jump, or run. They can get very active, so you need to make room for digging, as well as protection from predators and extreme weather changes. The space needs to be well-ventilated, thoroughly dry, and damp-free. A dirty environment causes illness and uneasiness. The height of that space also needs to be considered. You don't want their ears or heads touching the roof when they stand upright. You can use specific materials to build an effective shielded and cool area for them. Rabbits need a place to hide if they sense the presence of prey, for example, snakes, dogs, foxes, woodpeckers, and cats.

Rabbits need enough room for a hop, jump, or run.
https://unsplash.com/photos/ygqaLPkaB2o

For this purpose, you create some cozy holes in the rabbit room where they can escape when frightened. Rabbits can easily get bored and suffer when left in the same place for too long. This requires you to give them regular exercise. You may wonder what this consists of. Just let them out of their rooms into a protected space where they can hop and jump freely. One other vital element is the bed area. Rabbits can easily adapt to a cold temperature, but that may take time if they have never been exposed to it before. In the meantime, provide dust-free hay or straw, which is also safe to eat if they decide to take a snack at any time.

Selecting the Appropriate Hutch Materials

A hutch is a nesting area built specifically for raising rabbits. As a rabbit farmer, selecting suitable materials for building a hutch helps promote a healthy environment for your rabbit's growth. There are many elements for the perfect space, size, and design for raising your rabbits.

Hutch Size

Your hutch needs to be as big as possible. The minimum size for a rabbit breeding area should be no less than 12 square feet. You can add additional space for exercise purposes. The hutch should be designed so that the bed and exercise spaces are built together in one location so you don't have to move them to rest, feed, exercise, or excrete. As for the minimum spacing of a rabbit room, ensure that it is three to four times the size of the rabbit. You also need to consider the number of rabbits you'll be housing.

The greater the number, the larger the space needed. So, think upscale and future enlargement when planning.

Hutch Location

The hutch can be either indoors or outdoors. Look around your home and find unused cupboard areas or alcoves to transform into rabbit housing. Make sure you calculate the maximum number of rabbits likely to be contained in the housing being planned.

Design Materials

You must look for chewable and non-toxic materials when building your rabbit housing. Rabbits like to get busy, especially with their mouths, so you want to ensure that any materials found within their reach should at least be chewable without causing harm to them. There are so many options you can look out for when considering a decent house for your bunnies.

- **Woods:** Try out pine or plywood, which are common and preferable for outdoor housing. Even when ingested, they are harmless compared to an MDF, which is toxic. There are other textures of wood you can use. For example, trim can be used as a good cover for the edges of the house, which are likely to get chewed by the bunnies. For example, try a skirting board.
- **Plastics:** Plastics can hardly be avoided. A great blend would be using Correx, but it is not chewable. If you're still at the beginning of construction, using little or no plastic for housing is much preferable as, when broken, it can become sharp.
- **A Mesh of Wires:** To add to the aesthetics, you should add mesh. Mesh does not entirely need to be made of wire; you can finish up with something as simple as chicken wire or, preferably, a mesh coated with powder, plastic, or plastikote.

The latter are non-toxic when dried, come in different colors and look good. You can place it within the frame of doors and windows for a better fit to prevent chewing.

- **The Flooring:** Choosing the appropriate flooring that makes it easy for you to clean is your best option. Hard flooring is a much better option, for example, safety flooring or lino. They are both cheap and easy to install. A safety flooring is way harder in texture than a standard lino. You mostly find them in a vet's waiting room. To fit it, apply a flooring adhesive or a double-sided tape and lay it on top. To complete, you can apply a sealant around the edges to create the perfect finishing touches. Tiles are good options as well. The only catch would be that you'll need to avoid shiny and slippery tiles so your rabbits find it easy to move around. These floors should be easy to clean.

Determining Adequate Spacing for Each Rabbit

You can plan many elements to include in your rabbit enclosure, but it's easy to overlook space planning. It's easy when planning for one rabbit, but deciding on more isn't always that easy. You may under-calculate the space and end up packing the poor bunnies into a space that is too small. The minimum spacing required for a rabbit is dependent on many factors:

- Breed types
- Size of rabbit
- Weight of rabbit

You calculate the cage size by multiplying the length and width of the cage. Remember that all amenities inside the cage, including the water basin and food tray, must be subtracted from the result. When you provide the appropriate house space for your rabbits, you're guaranteed that they will grow and develop healthily.

How Much House Spacing Do Rabbits Need?

The home should be comfortable and easy to move around. However, rabbits' sizes vary, which should be considered when planning the housing size. They also vary in weight. For example, the Netherland dwarfs weigh just two pounds compared to the Flemish giant of 15 pounds. Make sure you take this factor into account. If your rabbit is still in its early growth stage, you will need to adjust your calculations for its possible adult size. If you're unsure of the eventual size, you can wait it out before opting for an enclosure expansion.

- **Minimum Hutch Length**

The best and easiest way to determine the hop length of your rabbit is by taking a measurement starting from their nose to toe when they are stretched out, then multiplying this length by three. This will give you the minimum enclosure length. For example, say you measure a small-sized three-pound rabbit while it was lying across the floor to be 12 inches long; 12 inches multiplied by three would give you 36 inches. This results in a 3-4 feet length enclosure, which should be the minimum. It should never be smaller than this because the rabbit only grows bigger and longer. They will feel cramped if you do not expand the enclosure with time.

- **Minimum Hutch Height**

As much as your rabbit's house length needs some calculating, so does the height. You don't want to create a compartment where they could end up hurting their heads when they hop. When there isn't enough room for them to stand on their feet, they can develop a spinal deformity. The worst case is that they may lose their spinal flexibility, a far greater risk for them. Giving your rabbits at least 2-3 feet of vertical spacing is as vital as horizontal spacing.

- **Minimum Width**

The width needs to be just as wide as the length, even more, to prevent any cramped conditions. To measure, you'll need to add a few more inches to your rabbit's already measured length, given from previous examples. Here's another example of how to apply this. Assuming the length of your rabbit is 14 inches or 16 inches at most, you would need to provide enough wiggle room that should be about 4×2 feet, which results in eight square feet for your rabbit.

Giving Your Rabbits Sufficient Space to Multiply

Are you thinking of increasing the population of your bunnies? Then, you'll need to put some more elements in place, like more bed space, food supply, and more room for exercise. Thinking about more spacing may seem overwhelming, but you'll have to start somewhere. For example, the space increment can depend on the size and weight of your rabbits. That's why having a playpen is a better bet, as it would save you the stress of immediate expansion of housing. A playpen can hold two bunnies with very low weight and size compared to bigger and heavier breeds. In this case, you need an expansion as quickly as possible. Rabbits grow within the blink of an eye.

Rabbits grow within the blink of an eye.
https://unsplash.com/photos/J_cqfq9FjmU

There isn't a specific guideline for knowing the right timing or calculation to expand your housing, but it's vital to understand that each female rabbit can give birth to four to eight little bunnies. When it eventually gets to this state, you might have to vacate your family room so they can all feel at home and happy. The next question would be, "How Can I Eventually Expand the Housing?" To begin, you must have a small enclosure sufficient for at least a rabbit. Then, you can start to expand the enclosures. If you did not have a rabbit house before, then it's even easier for you. All you'll have to do is follow the guidelines above. This will give you the perfect square feet for a healthy space. If this is not the case for you, don't worry. You can still expand spacing to give you your desired result. Here's how:

- **Expand Enclosure:** First, you must expand the size of your enclosure. To do this, you can attach other compartments, for example, an exercise pen. You can try this out even without completely replacing their home.

- **Utilize Spaces under Furniture:** If you live in a tiny home or apartment, you can use the space underneath your furniture for the rabbit's enclosure, creating more vertical spacing to compensate for the small length of the area. A good example of furniture to utilize would be under your dining table.
- **More Vertical Space:** If your rabbits are likely to hop and jump around frequently, then it would be wise to create more vertical platforms to give them more room to move around.
- **Roaming about Freely:** You won't need to contain your rabbit for longer if you feel they're sufficiently trained. You can seal off every exit to prevent escape and let them move around a bit.

Creating Space for Feeding, Nesting, and Waste Management

A rabbit's home is its environment. This is much more than where it eats, sleeps, or exercises. Wherever and whatever it can access can be classified as a homely environment. As mentioned earlier, this also contains the necessary amenities for its survival, including bedding, trays for food, hay, or straw. There should be adequate ventilation and protection from predators. A suitable resting rabbit home should contain at least 50% of the following:

- Undisturbed food and water.
- A place for resting and comfort.
- A place to exercise and explore safely.
- A place to hide when frightened.
- A space to chew whatever and whenever.
- A place to escape to interact with companions.
- A place for shelter from any change in temperature.

A rabbit's resting area could be expanded to different segments. As mentioned earlier, one would be a dark-covered room for sleeping away from the noise and the other for eating and relaxing. All spaces should be drought and free of dampness to avoid poor ventilation.

Housing and Waste Management

What is a suitable toilet area for your rabbits? Rabbits need access to a regular toilet site. For this purpose, you can provide trays lined with straw, hay, or newspapers. A rabbit's mouth is constantly busy with food, so you can be rest assured that they will pass a lot of waste. Ensure that the toilet area is separate from the sleeping area. Hay and trays used in the toilet areas must not be made of toxic material. In addition, making good use of wire for housing is essential, and applying solid flooring for easy and regular clean-up.

The bedding area should come with an extra insulator for extra cold weather. It is not advisable to use wood shelving as a bedding material, and the exercise area cannot be overlooked. There are still elements that need to be mentioned. For example:

- Rabbits should have access to a place to run daily.
- The exercise area must contain raised spaces for jumping. This room should be outdoors.
- It should be secured enough to prevent any entry of predators.
- If possible, it can be moved from time to time to avoid burrowing or overgrazing.
- Provide a covering or shade for windy or rainy days.
- There should be enough room for all the rabbits to be together or alone in one place.

Tips and Strategies for Maintaining the Hutch

Cleanliness

Hygiene is necessary for the health of your rabbits. You want to maintain the state and environment in which they live to prevent illness or disease. Here are a few factors to help:

- The rabbit sleeping area should be thoroughly cleaned on a daily basis. To do this, remove wet or dirty shelving or bed areas and take out spoilt or old food.
- The entire living area, indoors and outdoors, should be cleaned at least once every week. This must be done to maintain a clean, hygienic environment for your rabbits.

- Use mild pet-friendly disinfectant if possible. Just like a rabbit is fragile physically, their immunity is just as fragile.

Temperature Regulation
- Most healthy rabbits get acclimatized to an outdoor environment. They can withstand any temperature difference as long as they are provided with good nutritional food and housing.
- Those rabbits accustomed to indoor housing should not be suddenly placed outside in cold weather. If you want your bunny to stay alive during winter, you must gradually expose them to the indoors beforehand.
- Old or very young rabbits should never be allowed outside because they cannot tolerate vast temperature differences all at once.
- Certain temperatures are still considered too much for even a healthy adult rabbit, for example, 20 degrees Fahrenheit.

Creating a proper and healthy living environment for your bunnies requires study and measurements. If you want to raise a hutch of bunnies for domestic purposes, you must put certain factors into place. Rabbit farming can be fruitful enough to provide a profitable and resourceful income for a farmer. It is also a good source of quality protein meat. The art of raising rabbits for meat is known as Cuniculture, and to begin farming, you need a good location with a good grazing source, preferably far from a residential area but close enough to a commercial setting. You don't have to start on a large scale. You can begin small and gradually build your numbers from there.

Choose a location with adequate transportation close by. Apart from this, your rabbits must be in good health, so creating enough room for growth and exercise, where temperature change can easily be regulated, is necessary. With all this in place, you are well on your way to a profitable rabbit farming business.

Chapter 4: Understanding Your Rabbits' Nutritional Needs

If you've gone down the rabbit hole of looking for the perfect rabbit meal plan, you're now probably confused and a little lost, too. Some people claim that vegetables and greens are all rabbits need to survive, while others advise against feeding rabbits too much of the green stuff. Then there are the people who swear by pellet feedings and consider pellets to be the only answer to rabbit sustenance. Hay is also considered a suitable food source for your rabbits. All these opinions and options are enough to make anyone feel confused.

In the midst of this nutritional overload, you find yourself playing the role of a rabbit diet detective, carefully calculating the perfect balance of greens, pellets, and hay. However, the thing is, there's no perfect meal plan that suits the needs of every bunny, though there is a general selection of nutritional foods you can follow. Although each bunny has its own preferences, some diet options are preferred over others for meat rabbits.

Types of Feed for Rabbits

When you're thinking about what to feed your rabbits, it's a good idea to consider your expertise. Suppose you're not an expert in nutrition, or you don't want to dive into the intricacies of ration formulation. In that case, there's a simple route to take. Start off with commercial rabbit pellets, as they're a reliable standard meal. As you get more comfortable, you can try the other feeding options. Remember, when it comes to feeding your rabbits, doing your homework is a must. Don't assume that because wild rabbits munch on grass, your meat rabbits can survive on the same menu. Don't just feed them anything because some vegetables can be *very toxic* for them.

On the other hand, the cost of rabbit food keeps climbing (and there's no sign of it slowing down anytime soon), which may make the idea of growing your own rabbit food sound appealing. However, only try this if you can be consistent with the process. Otherwise, it's all just a whole lot of waste.

Although each bunny has its own preferences, some diet options are preferred over others for meat rabbits.
https://unsplash.com/photos/sodF0c8xm-0

1. Pellets

If you're looking for a quick, easy, and well-balanced option, go for good-quality organic or non-organic pellets, as they're a safe bet. If you've ever thought about making your own rabbit food from scratch, be warned that it's not as straightforward as it may seem, given all the different factors you need to consider. If you're after an uncomplicated route, pellets are the solution. Eventually, you can gradually introduce fresh foods into the mix. In this way, you could even save some money by blending pellets with weeds or adding extra greens from your garden. If you want to rear your bunnies on a primarily fresh greens diet, then make sure the breed you choose can handle this kind of diet. Alternatively, you could connect with someone who's already raising rabbits and loves their leafy meals. Pellets are like a rabbit's dream meal because they're perfectly formulated and balanced to meet all their nutritional needs. It's a tailor-made feast of essential vitamins and minerals, all designed to keep rabbits healthy and happy.

Alternatives to Pelleted Feed

You can also feed rabbits with what you grow in your garden or gather from the pasture. The joy of knowing you're providing something entirely homegrown can be very satisfying. However, it's essential to make sure your rabbits get a well-rounded diet for their well-being. Given the growing fascination with steering clear of pellet feeds and the move toward choosing natural, homegrown foods for your meat rabbits, here are some feed alternatives suitable for them:

1. Hay

Rabbits need high fiber for most of their diet. This can be met by feeding them hay. The basic ingredient of their meals should be quality grass hay. Look for hay that's clean, free of dust and mold, and which has enough protein to keep their systems working smoothly. Grass hay is the best choice. It's packed with fiber, which does wonders for their digestion. Make sure you avoid straight alfalfa hay, though many people think otherwise. For instance, alfalfa isn't grass; it's a legume that's fed to animals to increase their protein intake. Although plant protein is considered good for rabbits, alfalfa contains excessive calcium, which isn't favorable for your rabbits. In fact, it can result in concentrated urine, which leads to kidney stones, which is something you want to avoid under any circumstances. You can also try some other hay options, like Timothy grass or high-quality horse hay. If you wish to use alfalfa, you can combine it with grass to balance the nutrients. Oat grass is a good option for this, which can be easily found at horse feed supply stores.

2. Greens

Most cartoons show rabbits eating carrots and other vegetables, but did you know that a lot of these greens are not good for your rabbits? In fact, there are some greens you should steer clear of. For instance, Iceberg lettuce, although rabbits love to eat it, is, in fact, toxic for them. It's too watery and can lead to upset tummies and messy stools. Instead, go for dark and leafy greens like kale and leaf lettuce. These are bursting with vitamin A and other nutrients. A tip to remember is that once greens start looking old, they can become a fermenting mess. Stick to the fresh ones and offer only what your bunny can finish in around 15 minutes. Other greens that are suitable for your rabbits include radish greens, sunflower leaves, beet greens and roots, carrot tops, dill, mint, comfrey, and more. These are the greens your rabbits can happily eat.

3. Treat Foods

Treats like carrots, fruit, and starchy foods are packed with sugar. You should always give these treats in very small amounts. Why the caution? Well, the high sugar levels can really mess with your rabbit's gut health and cause digestive problems.

4. Rabbit Food Caveats

You should know that wild rabbits can munch on almost anything. However, your pet rabbits, or even rabbits grown for meat, cannot do the same. They belong to different species. While they do have some common dietary preferences, they don't necessarily eat the same things. When wild rabbits are out and about, they nibble on fresh forage right where it grows. But it's not the same when it comes to your domestic rabbits. Dumping a bunch of restaurant veggie scraps into their pens is not a good idea. Even though your rabbit is an herbivore, this is a move you'll come to regret. Firstly, they won't get the proper nutrition, and secondly, those scraps will end up wilting and fermenting on the pen floor, drawing flies and causing a mess. And those yard clippings you had planned to give them are too delicate and will have already started wilting by the time they reach your bunny's bowl. Remember, what works for wild rabbits isn't always fit for their domestic counterparts.

5. Alfalfa Hay and Rolled Oats

A simple alternative feed idea is to use a mix of alfalfa hay and rolled oats. Rabbits usually love this combination and prefer it over regular hay. However, as already discussed, alfalfa hay has a considerable amount of protein and calcium. So, if you decide to go the alfalfa route, you've got to team it up with rolled oats so that your rabbits get some extra phosphorus, which helps balance out the high calcium levels from the alfalfa.

6. Oats and/or Barley

Consider oats and/or barley as a solid option. They work wonders, especially for those growing kits who are just starting to explore the world of food beyond milk. For the little ones who've recently weaned, these grains are gentle on their tummies and easy to digest. A good option is to keep a separate bowl of oats right inside the cage for the young ones. When choosing the oats and barley categories, opt for the uncut and unrolled options, as they're the best suited for beginners.

7. Black Oil Sunflower Seeds

You'll find these seeds in the bird feed section, and although they're typically used to feed birds, they work like magic on rabbit coats, too. If you're looking to give your rabbits a stunning appearance, consider giving them about a teaspoon of BOSS every day.

8. Alfalfa or Hay Cubes

Instead of giving your rabbits rough hay, try hay cubes. These neat little compressed blocks are made from alfalfa or hay. They're not just plain cubes; they're also infused with molasses and packed tightly. They're like a chewy treat your rabbits can gnaw on, and that's important because rabbits' teeth never stop growing. You can find bags of these cubes in a rabbit feed store, or if you prefer the bigger ones, you can always head over to the horse feed store or section. The latter actually provide better value and are also helpful for your rabbit's dental health.

9. Calf Manna

This is something that stands out from the crowd. Calf Manna is not just a name; it's a brand of supplement that works wonders. This is specially crafted to boost milk production in a variety of animals. If you have a pregnant or nursing rabbit, giving her a couple of teaspoons of Calf Manna each day can make a big difference. It's especially good for meat-breed rabbits that tend to have quite sizable litter. By ensuring your mama rabbit gets her Calf Manna, you're helping her provide enough milk for her kits and ensuring she stays in tip-top shape throughout pregnancy and nursing. This smart move could even allow you to breed her sooner for another round of kits.

10. Dried or Fresh Fruit

Both dried and fresh varieties of fruit are great for rabbits. These colorful treats can add a little excitement to your rabbit's diet. However, while they're a nice occasional treat, it's important not to overdo it. In addition, many of these fruits deal with specific problems. For example, pineapples can actually help if your rabbit suffers from a bout of "fur block." This happens when rabbits ingest too much of their own fur, causing a blockage in their digestive system. And then there's papaya - not only is it tasty, but it can also serve a practical purpose. If you notice your rabbit's urine has a strong odor, papaya can help reduce that.

11. Weeds, Lawn Trimmings, and Bush Trimmings

Some natural feed for your rabbits includes weeds, lawn trimmings, and bush clippings. These can actually be pretty useful for your rabbits. Greens not only include the vegetables you can feed your rabbits but also grass, weeds, lawn trimmings, and even leaves. Just be sure they're on the safe-to-eat list. Some of the good wild plant choices are comfrey, chickweed, cow parsley, docks, cattails, dandelion, plantain, shepherd's purse, sow thistle, and watercress. You might want to refer to a safe list online.

Dandelions are like rabbit candy – they love them so much that you may just become a dandelion farmer in your own yard. Freshly cut grass is another winner in their eyes. Many people set up a little rabbit play area with wire fencing or use a dog crate to let their bunnies roam and nibble on these natural goodies while they tidy up their living spaces. It's a win-win situation, but be extra careful to make sure there are no toxic weeds within reach.

Nutrient Breakdown

Your rabbit's diet should have a combination of nutrients to ensure they can grow most effectively.

- **Carbohydrates**

Think of these as energy boosters. Rabbits can balance their own diet – they'll munch more if their energy level is low and less if it's high, but too much energy (read: carbs) can actually slow down their digestion. So, tread carefully and find the right balance.

- **Fiber**

Fiber is a rabbit's best friend. Wild rabbits eat loads of it, and while young bunnies need a bit less, it's still super important. When feeding your adult rabbits, it's best to have at least 25% fiber in their food. So, look for the ones with the highest fiber content.

- **Minerals**

The rabbit food discussed above usually contains all the minerals required for a healthy diet except for cobalt. It is the missing component that you have to fill some other way.

- **Vitamins**

Your rabbits have some friendly bacteria in their intestines - vitamin B-complex and vitamin C, which also means they need to get vitamins A, D, and E from their diet. So, make sure these vitamins are included in their pellet mix.

Keep in mind that moderation is key. Don't go overboard with food. Feed your rabbits about twice a day to keep their intake balanced. However, stay clear of fermented and sour foods, as they can create problems. If you prefer to use pellet food for your bunnies, keep an eye on their weight because they can gain too much too fast, which is something you'll want to avoid.

Balancing the Diet

Rabbits have unique dietary needs, and getting the right balance of nutrients is vital to keeping them flourishing. The best way to achieve this is to combine all the different food sources. You can start with commercially produced rabbit feed, i.e., pellets. These specially formulated feeds are a nutritional goldmine designed to meet your rabbits' dietary requirements. While the idea of crafting your own mix is tempting, it's a bit of a nutritional tightrope, and finding that perfect balance can be challenging. So, leaning on the expertise of commercial feeds is a smart choice.

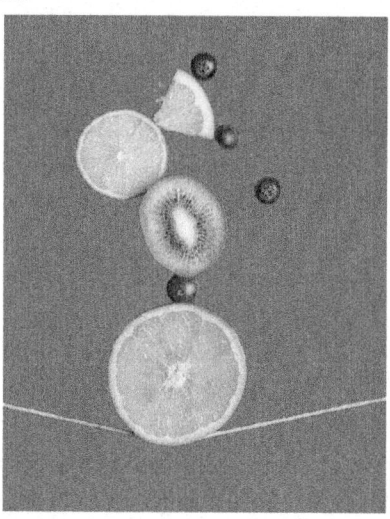

Rabbits have unique dietary needs, and getting the right balance of nutrients is key to keeping them flourishing.
https://www.pexels.com/photo/fruit-slices-balancing-on-a-line-7465042/

Next, there's protein – a key player in your rabbit's diet. Commercial feeds usually offer protein levels ranging from 14 to 18 percent. For rabbits raised with a focus on meat production, a diet rich in protein (around 16 to 18 percent) can be a growth accelerator. Keep your feed cool and dry to prevent unwanted mold growth. The airflow should be suitable, and make sure not to leave the feed open and accessible to sneaky rodents. Keep it protected in chew-proof containers.

In addition, you should incorporate hay into your rabbit's diet, as it not only supplements their diet but also keeps them engaged and helps maintain their dental health. There's a variety of hay types to choose from; your choice of hay should align with your rabbits' dietary needs. For example, if you're providing lower-protein pellets, consider balancing it out with high-protein alfalfa hay.

Feeding Guidelines

The amount of food your rabbits need isn't a one-size-fits-all deal. Each rabbit has different nutritional needs, especially when characterized as young or adult. It also depends on the rabbit's living conditions and how much food they should be given. For instance, when the weather turns cold, they need a bit more food, whereas in the summer, they can be given less.

You can also decide how much food to give to your rabbits by keeping an eye on their weight. If they look too lean, they need more food, and vice versa. It's wise to keep a watchful eye on their portions when it comes to full-grown bucks that aren't currently breeding. The goal is to avoid pudgy rabbits – too much fluffiness can mess with their fertility and turn them into couch potatoes. On average, adult rabbits munch about four ounces of food per day. If you have rabbits with little ones, they'll need about eight ounces to keep up with the parenting hustle.

For those meaty breeds, the food scoop ranges from 1/2 to 1 cup daily – but it varies from rabbit to rabbit, just like your own food preferences. Now, here's where the debate comes in. Pregnant or nursing rabbits and growing kits may enjoy the luxury of free-feeding, and breeders will either nod in agreement or share raised eyebrows. More protein often translates to speedier growth and bigger bunnies, but the free-feeding question is a balancing act.

What to Do if a Rabbit Is off Their Feed

When your rabbit's appetite takes a detour, your first instinct may be to offer up some familiar treats or greens they've enjoyed in the past. It seems like a quick fix to get them nibbling again. However, these treats can sometimes fuel the fire, stirring up trouble in their digestive system and leading to loose stools. Instead, you should use a different approach. Offer a helping of good, clean grass hay, as it's like a soothing balm for their upset tummies. Another option on the menu is rolled oats, a high-fiber choice that's gentle on the digestive tract and a treat for their taste buds.

Speaking of essentials, don't forget the water supply. Rabbits need fresh, clean water. Give the water bottle or water line a quick glance to clear any blockages. Rabbits can get a bit finicky, and they're not ones to wait around for a drink. Without water, they can quickly dehydrate, and that's not good for their appetite. If your rabbit still isn't eating, check their droppings. If things are looking a bit runny, their diet needs more high-fiber food.

In conclusion, the diet of a rabbit controls many things, especially when you're growing them for meat. This dietary balance isn't just about filling bellies. It's a requirement for their growth and their ability to breed. Think of it as a delicate balance where the right mix of nutrients fuels their development, ensuring those young kits reach their full potential. And it's not just about the physical gains. The rabbit's diet can influence their behavior, energy levels, and even their reproductive abilities. Keep in mind that whenever you're tinkering with your rabbits' menu, take it slowly. Quick changes can spell trouble, and you definitely don't want that. Remember to give them plenty of water - not just a sip – as they need plenty of it.

Chapter 5: Preventing and Addressing Health Issues

Rabbits can develop health issues, diseases, and disorders just as easily as any other animal. When raising rabbits for meat, taking care that your bevy of bunnies is healthy and thriving is part of the deal. Knowing about common diseases and disorders can make it easier to address these health issues and prevent recurrences. This chapter provides an overview of the common diseases rabbits can develop, guidelines for prevention, treatment protocols, and everything else you need to know to keep rabbits healthy.

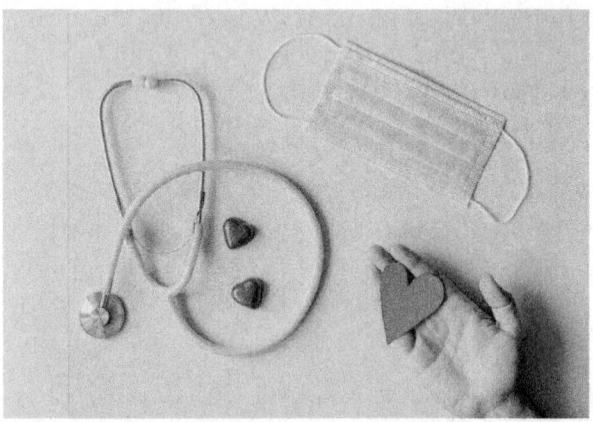

When raising rabbits for meat, taking care that your bevy of bunnies is healthy and thriving is part of the deal.
https://www.pexels.com/photo/medical-stethoscope-and-mask-composed-with-red-foiled-chocolate-hearts-4386466/

Common Health Issues Rabbits Face

A rabbit can live and reproduce for at least eight years when fed and cared for properly. However, there are several common diseases that rabbits can develop as they age. Here are some common conditions you need to be familiar with.

Respiratory Tract Infections

Unlike humans, a rabbit can only breathe through their noses. As the nose is the only orifice in rabbits for breathing, air-borne microorganisms and harmful chemicals can easily get in and infect the respiratory system. Although they have an immune system that can ward off harmful organisms and break down toxic chemicals, severe or prolonged exposure can eventually lead to an infection.

Rabbits with respiratory illnesses will sneeze repeatedly, and their breathing will be labored. These signs are associated with upper respiratory infections. Lower respiratory infections also occur where an added wheezing sound can be heard when you listen up close.

GI Stasis (Gastrointestinal Stasis)

GI stasis occurs when a rabbit's digestive system stops working or slows down, disrupting the normal movement of food and waste through the gut. This can be caused by a diet low in fiber, dehydration, stress, or other underlying health issues. Without proper food movement, the gut becomes compacted and leads to painful gas buildup, bloating, and discomfort. Symptoms include reduced appetite, smaller or no fecal pellets, lethargy, a hunched posture, and sometimes a visibly distended belly. GI stasis can be serious and even fatal if not attended to promptly. Watch for changes in appetite. A sudden decrease in eating or reluctance to eat hay and fresh vegetables can be a sign.

The two pivotal measures you need to keep an eye on include monitoring fecal output, looking for smaller, fewer, or abnormally shaped fecal pellets, and observing posture. A hunched posture or sitting in a stretched-out position is a sign of GI.

Dental Problems

Rabbits' teeth grow continuously, and if they become misaligned or overgrown, it can lead to various dental issues. Overgrown teeth can cause pain, injury to the cheeks and tongue, difficulty eating, and weight loss. Dental spurs, sharp points that develop on the teeth, can also cause discomfort. These problems often stem from genetics or an improper diet lacking sufficient fiber to wear down the teeth naturally. If your rabbit is dropping food, chewing with one side of the mouth, or avoiding certain foods, it could indicate dental issues. Likewise, excessive salivation can be a sign of mouth pain.

Respiratory Infections

Respiratory infections are caused by bacteria such as Pasteurella multocida. The symptoms are nasal discharge, sneezing, coughing, labored breathing, and conjunctivitis (inflammation of the eye lining). Stress, poor ventilation, and overcrowded living conditions can increase the likelihood of respiratory issues. During regular monitoring, observe your rabbit's breathing pattern. Rapid, labored, or noisy breathing can indicate a respiratory issue. In the case of a respiratory infection, you will also see clear or clouded discharge from the nose.

Pasteurellosis

Pasteurellosis is caused by the bacterium Pasteurella multocida. It often manifests as upper respiratory infections with symptoms like sneezing, nasal discharge, and eye discharge. However, it can also lead to more severe conditions such as abscesses (localized pus-filled swellings), particularly around the head and neck area. Rabbits with weakened immune systems are more susceptible to pasteurellosis. Always check for swelling around the head and neck region that can indicate the presence of abscesses.

Ear Mites

Ear mites are tiny parasites that infest a rabbit's ears, causing irritation, itching, and inflammation. Rabbits with ear mites may scratch their ears excessively, tilt their heads, and show signs of discomfort. Left untreated, ear mite infestations can lead to secondary bacterial infections and ear hematomas (blood-filled swelling in the ear flap). Regularly check the rabbits for ear scratching, redness, and swelling around the area covering the ears.

Myxomatosis

Myxomatosis is a viral disease that's usually spread by biting insects. It causes swelling and discharge around the eyes, ears, and genitalia. The virus weakens the rabbit's immune system, leaving them vulnerable to secondary bacterial infections. The disease progresses quickly and can be fatal within a week or two. Look for facial swelling where swollen eyes, ears, and face are characteristic signs. Furthermore, look out for watery or pus-like discharge from the eyes, nose, or genitalia.

Rabbit Hemorrhagic Disease (RHD)

RHD is a highly contagious viral disease that primarily affects the liver and blood vessels. It can lead to sudden death or internal bleeding, causing bloody discharge from the nose, mouth, or rectum. There are different strains of RHD, and its severity can vary. This disease poses a significant risk to unvaccinated rabbits. Watch for sudden death, as rabbits affected by RHD can die suddenly with no signs of disease. You should also look for bleeding from the nose, mouth, or rectum, and if you notice bloody discharge, immediately seek veterinary help.

E. cuniculi Infection

Encephalitozoon cuniculi is a microorganism that causes neurological issues in rabbits. It often affects the brain and kidneys. Infected rabbits will show symptoms such as a tilted head, seizures, incoordination, and urinary problems. The infection can be difficult to treat, leading to more chronic health problems. In case of an infection, there will be persistent head tilt or circling and difficulty walking.

Uterine Tumors

Female rabbits that haven't been spayed are at risk of developing uterine tumors, particularly adenocarcinomas. These tumors cause hormonal imbalances, uterine infections (pyometra), and pain. Spaying female rabbits at a young age will significantly reduce the risk of uterine problems. Swelling in the abdominal area and teeth grinding are some common signs that indicate uterine issues in female rabbits.

Skin Conditions

Rabbits can develop various skin issues, such as fur mites that cause itching and hair loss. Ringworm, a fungal infection, leads to circular areas of hair loss and skin inflammation. Abscesses are pus-filled swellings that occur anywhere on the body and are often caused by bacterial infections. Missing patches of hair and frequent scratching indicate the possibility of an underlying skin infection.

Obesity

Overfeeding and a diet high in carbohydrates leads to obesity. Obesity can result in joint problems, respiratory difficulties, and a reduced quality of life. It's essential to monitor a rabbit's diet and provide plenty of opportunities for exercise. Regularly assess your rabbit's body shape and weight. Overweight rabbits often have a round, bulging appearance.

Preventative Care Tips

Preventive care is essential to keep rabbits healthy and minimize the risk of common illnesses. Here are some vital tips for providing the best preventive care for your rabbit.

Proper Diet

Your rabbits should be provided with fresh, hygienic, and quality foods compatible with their tummies. Some recommendations include:

Hay

Rabbits should have access to high-quality grass at all times. Common hay types best for rabbits are timothy, meadow, and orchard grass. The essential fiber in the hay promotes health by helping rabbits digest food better.

Rabbits should have access to high-quality grass at all times.
https://www.pexels.com/photo/agriculture-arable-bale-countryside-289334/

Fresh Vegetables

Offer a variety of fresh, rabbit-safe veggies daily, like leafy greens (kale, romaine, parsley) and limited amounts of other veggies like carrots and bell peppers.

Limited Pellets

Although you can find several pellet varieties on the market, these pallets can never be used to substitute other foods. You can feed rabbits pallets that are high in fiber and low in calcium in limited quantities.

Water

Rabbits should always have access to fresh and clean water at all times. Not keeping the water clean increases the chances of transmittable diseases in the warren.

Regular Exercise

Allow your rabbits to have safe access to a larger space to exercise and explore, such as in a rabbit-proofed room or an exercise pen. You can provide toys like cardboard boxes, tunnels, and safe chew toys to keep your rabbits mentally and physically active.

Hygienic Housing Conditions

Clean your rabbit's living space regularly to prevent the buildup of waste and harmful microorganisms. A clean environment promotes good health and prevents the development and transmission of diseases. Surprisingly, your rabbits can also be litter trained, which makes cleanup easier. Depending on the breed, rabbits may need regular brushing to prevent matting and to remove loose fur. Densley-coated rabbit species with long fur will require daily grooming, whereas rabbits with short hair don't need much attention. Lastly, at least once a month, trim your rabbit's nails to prevent overgrowth and discomfort. Be careful never to cut below the quick.

Regular Vet Check-Ups

Find a veterinarian experienced with rabbits and schedule regular check-ups to catch any health issues early. Keep vaccinations up to date, including those for rabbit hemorrhagic disease (RHD) and myxomatosis.

Rabbit-Proofing

Make your home safe by securing cords, removing toxic plants, and blocking off access to dangerous areas.

Social Interaction

Rabbits are social animals. Spend time interacting with your rabbit daily to provide mental stimulation and companionship.

Weight Management

Monitor your rabbits' weight and body condition to prevent obesity. Adjust their diet and exercise routine accordingly.

Avoid Stress

Minimize stressors such as sudden changes in the environment, loud noises, or aggressive handling.

Parasite Prevention

Follow your vet's recommendations to prevent external parasites like fleas and mites.

Quarantine for New Additions

If you're introducing a new rabbit, quarantine them for a few weeks before introducing them to your existing rabbit(s) to prevent potential disease transmission.

Remember that rabbits have unique needs, and it's important to stay informed and educated about their care. Providing a well-rounded, balanced lifestyle with proper diet, exercise, hygiene, and medical attention will go a long way toward ensuring your rabbit's health and happiness.

Seeking Care

Knowing when to seek professional veterinary help is crucial for their well-being. If you notice any unusual behavior, symptoms, or changes in your rabbit's condition, it's best to consult a veterinarian experienced in rabbit care. Here are some guidelines on when to seek veterinary assistance and the types of treatments that might be necessary.

Emergency Situations

Seek immediate veterinary help if you observe any of the following:

- Severe difficulty breathing or gasping for air
- Sudden lethargy, weakness, or collapse
- Profuse bleeding from any part of the body
- Seizures or severe head tilt
- Distended belly, especially if accompanied by pain and discomfort
- Uncontrollable diarrhea or constipation
- Severe trauma or injury

Behavioral Changes

Rabbits are experts at hiding signs of illness. If you notice changes in behavior or routine, it could indicate a health problem:

- Decreased appetite or refusal to eat
- Reduced water intake

- Lethargy and reduced activity
- Isolation and hiding more than usual
- Teeth grinding (a sign of pain)
- Aggressive behavior or changes in social interaction

Gastrointestinal Issues

GI stasis, diarrhea, or constipation are common rabbit health issues.

- Seek help if your rabbit has not eaten or produced any fecal pellets for more than 12 hours.
- If your rabbit's stools are consistently soft or watery, or if they have difficulty passing stools.

Respiratory Symptoms

- Nasal discharge, sneezing, wheezing, labored breathing, and coughing can be signs of respiratory infections.
- If your rabbit is having trouble breathing or has a visible discharge, contact a vet.

Dental Problems

- If your rabbit is drooling, pawing at its mouth, or showing reluctance to eat, dental issues might be the cause.
- Overgrown teeth or dental spurs require professional trimming by a veterinarian.

Skin and Fur Issues

- Scratching, hair loss, scabs, or skin lesions could indicate mites, ringworms, or other skin conditions.
- Abscesses, lumps, or unusual growths should always be reported to a vet and examined.

Eye and Ear Issues

- Cloudy or bulging eyes, excessive tearing, redness, or discharge warrant veterinary attention.
- Head tilt, circling, and balance problems may indicate an inner ear infection or E. cuniculi infection.

Reproductive and Urogenital Issues

If you have an unspayed female rabbit, watch for signs of uterine issues like bleeding, swelling, or discomfort.

Male rabbits with difficulty urinating or producing urine could have urinary tract problems.

Vaccinations

Consult your vet about the recommended vaccination schedule for diseases like rabbit hemorrhagic disease (RHD) and myxomatosis. Treatments for rabbit health issues can vary widely and should always be determined by a veterinarian. They may include:

- Antibiotics or antiviral medications for infections
- Pain relief and anti-inflammatory medication
- Dental procedures for overgrown teeth or dental spurs
- Fluid therapy to combat dehydration
- Surgical procedures to remove abscesses or tumors
- Parasite treatment for external or internal parasites
- Supportive care such as syringe feeding, hydration, and temperature regulation

Vaccinations to Prevent Specific Diseases

The key is to seek professional veterinary care as soon as you notice any signs of illness or discomfort in your rabbit. Rabbits are delicate animals, and early intervention can make a significant difference in their prognosis and recovery.

Conduct in Emergency Situations

Stay Calm

In any emergency situation, staying calm is essential. Take a moment to collect yourself before taking action. Rabbits are sensitive to their owner's emotions, and your calm demeanor will help keep your rabbit from getting more stressed.

Assess the Situation

Quickly assess the situation to understand the severity of the emergency. Is your rabbit injured, showing signs of illness, or in immediate danger? This assessment will help you prioritize your actions.

Isolate and Protect

If your rabbit is in danger or is injured, gently move them to a safe and quiet area. Use a carrier or a confined space to prevent them from further harm or stress. Cover the carrier with a blanket to provide a sense of security.

Contact Your Veterinarian

Reach out to your veterinarian or an emergency veterinary clinic experienced with rabbits. Explain the situation and provide as much detail as possible about your rabbit's condition. Follow their advice and instructions closely.

First Aid

If your rabbit is bleeding or has an injury, use a clean, sterile cloth or gauze to apply gentle pressure to the affected area. Do not apply direct pressure to the eyes, nose, or mouth. Try to keep the area clean and minimize further trauma.

Breathing Difficulties

If your rabbit is having difficulty breathing, make sure they are in a well-ventilated area. Avoid drafts or extreme temperatures, as rabbits are sensitive to temperature changes. Keep them calm to reduce stress.

Stay Warm

In cases of shock or injury, your rabbit's body temperature can drop quickly. Cover them with a blanket or towel to help maintain their body heat. Be cautious not to overheat them, as rabbits can also become overheated.

Administer First Aid

Only administer first aid if you are trained – and it's safe. For example, if your rabbit is choking, clear their airways carefully. Always be gentle and avoid causing additional harm.

Transport to the Vet

If your veterinarian advises you to bring your rabbit in for immediate care, take them in as soon as possible. Secure a well-ventilated carrier in your vehicle to prevent sudden movements that could worsen their condition.

Keep Records

Document the symptoms you observed, the timeline of events, and any first aid you provided. These details will be valuable to the veterinarian in making an accurate diagnosis.

Follow Vet's Instructions

Follow your veterinarian's instructions carefully. They will guide you on how to stabilize your rabbit before you reach the clinic or tell you immediate steps to take.

Stay with Your Rabbit

If your rabbit requires hospitalization or treatment, stay in contact with the veterinary team. They will keep you informed about your rabbit's condition, treatment plan, and progress.

Remember, while you can provide some initial first aid, professional veterinary care is crucial for properly diagnosing and treating your rabbit's condition. Even if your rabbit seems to recover after first aid, it's still vital to seek professional assessment to ensure there are no hidden injuries or complications. Keeping your rabbit's safety and well-being a top priority in an emergency will help ensure the best possible outcome.

Owning a rabbit comes with the responsibility of safeguarding its health and well-being. These adorable creatures may be small, but they can be prone to health issues. By adopting a proactive approach to disease awareness and early management, you can ensure your rabbit enjoys a long and healthy life.

Understanding the Importance of Staying Informed

Being aware of common rabbit illnesses is essential. By understanding the signs and symptoms of conditions like gastrointestinal stasis, dental problems, and respiratory infections, you'll be better equipped to spot potential health concerns before they escalate.

Early Action

Recognizing warning signs is only the first step. Acting swiftly is crucial. If you notice any changes in your rabbit's behavior, appetite, or weight, don't hesitate to consult a veterinarian experienced in rabbit care. Your quick response can make all the difference in your rabbit's overall health.

Prioritizing Prevention

Preventing health issues is always preferable to treating them. Ensure your rabbit's diet is balanced and high in fiber, offer plenty of opportunities for exercise and mental stimulation, and keep its living space clean. Regular veterinary visits, vaccinations, and proactive grooming practices are also vital components of a preventive care routine.

Creating a Rabbit-Centric Lifestyle

Your rabbit's well-being should be at the center of your efforts. Make time to observe its behavior, engage in interactive play, and provide a comfortable and stress-free environment. By keeping your rabbit content and mentally stimulated, you contribute to its overall health.

Championing Your Rabbit's Health

As a responsible rabbit owner, you have the power to be your pet's health advocate. Taking a proactive approach to disease awareness and early management demonstrates your commitment to its happiness. Remember, your rabbit depends on you for its care. By prioritizing its health, you're ensuring that it enjoys a fulfilling life by your side.

Chapter 6: Ethical Breeding Practices for Sustainability

It can be tempting to just jump into raising rabbits and learn by going with the flow through trial and error. However, rabbits are living beings. Playing with the lives of the rabbits that will provide you with nourishment seems unnecessarily cruel. Therefore, ethical considerations must be taken into account when breeding rabbits. These considerations include health care, genetic diversity, housing, as well as understanding their reproductive cycles. By first gaining a deep understanding of the multiple factors that contribute to breeding healthy rabbits, you can make informed decisions to create the best environment for your rabbits. Sustainable backyard farming requires a conscious approach to animal rearing. Separating yourself from the often-cruel practices of large-scale industrial farming requires ethics to be applied to breeding and raising meat rabbits.

Rabbits are a lean source of protein, and they reproduce quickly. Setting up the frameworks to harness the fast rate of maturity in rabbits and the relatively cheap costs of raising them must align with a high ethical standard. Since you are raising your own meat, it is your responsibility to ensure that your animals live comfortably before slaughter. With a little bit of knowledge and by sticking to ethical principles, you can build a high-yield rabbit farm that takes up minimal space and is environmentally friendly. Once you have created your breeding system, it will be a lot easier to compassionately maintain a rabbit paradise. Therefore, exploring techniques to care for and breed rabbits ethically is central to an efficient backyard operation.

Managing Breeding Pairs

Understanding lineage is an essential component of ethical rabbit breeding. Ignoring genetic factors when breeding your animals could be disastrous and leave you with numerous defects and diseases. Choosing breeding pairs requires a basic comprehension of rabbit biology and their social behavior. Furthermore, there are many rabbit breeds to choose from, including the Flemish giant, the Californian, as well as the New Zealand White. Meat rabbits are chosen for their excellent bone-to-meat ratio, as well as their large size. It can be tempting to select the largest rabbits out of your litter for breeding to get more meat, but other factors determine which rabbits can be bred according to moral principles. A mindset driven by the quantity of meat production at the expense of quality could be a significant hindrance to crafting ethical backyard farming.

Understanding lineage is an essential component of ethical rabbit breeding.
https://www.pexels.com/photo/2-rabbits-eating-grass-at-daytime-331.52/

Many breeders are profit-driven, which can often result in undesirable treatment of rabbits. If you are planning to sell rabbit meat or even breed them for your own consumption, one of the worst approaches you can take, ethically speaking, is to look at rabbits as a simple product. The aim is to give rabbits the most comfortable life before they are ultimately slaughtered for meat. The relationship between a breeder and their rabbits should be mutually beneficial. The breeding decisions you make will determine how well your rabbits socialize, how healthy they will be, and, eventually, how much yield you will get from them. Therefore, managing breeding requires in-depth attention to the genetic diversity of rabbits, preventing in-breeding, as well as sustaining a healthy population size that your housing and space can accommodate.

Genetic Diversity

The genetic diversity of your rabbits will largely depend on the type of breeding that you pursue. The main types of breeding for rabbits are line breeding, crossbreeding, outcrossing, and inbreeding. Crossbreeding is when you select rabbits from completely different breeds that have various characteristics and mix them so that you can maximize genetic diversity. The issue with crossbreeding is that you cannot register the rabbits with the American Rabbit Breeders Association because they are not purebred. Crossbreeding will limit the possible buyers for your rabbits because purebreds are more desirable on the market. Outcrossing addresses the problem of crossbreeding, producing lines that are not purebred. Out-crossing is breeding rabbits of the same breed from different lineages. Line-breeding refers to breeding rabbits from the same family. However, the breeder is careful to make decisions that create genetic diversity by mating rabbits that have some familial distance. For example, rabbits bred with line-breeding may be half-siblings, or grandchildren will be bred with grandparents. Inbreeding refers to members of the same litter breeding.

Outcrossing is one of the best methods to ethically breed rabbits. Crossbreeding creates genetic diversity. However, it could cause problems with birthing. For example, if a larger male breed mates with a smaller female breed, it can cause issues in the birthing process because the resulting mix could mean that the kits are too large for the mother. Line-breeding creates more consistency amongst your rabbits, and you have more control over choosing desirable characteristics, but it can lead to genetic inferiority if the technique is used long-term. Inbred rabbits are genetically similar, so litter can often be prone to disease or have some physical defects.

Preventing Inbreeding

Keeping track of where your breeding lines come from is one of the first steps toward preventing inbreeding. A simple way you can ensure your rabbits are not directly related is by buying rabbits from different farms. Bodies like the American Rabbit Breeders Association exist in part to keep track of the lineages and breeds of different rabbits. Therefore, to prevent inbreeding, you must be aware of where your rabbits are obtained. Furthermore, it is better to get rabbits from registered breeders because there is a paper trail that you can follow to see precisely where the lineages originate and how pure (or not) they are.

Once you have acquired your breeding rabbits, the most effective way to prevent inbreeding is to be observant. Depending on the breeds you are raising, the age of sexual maturity will be different. Once a rabbit has reached sexual maturity, they should be kept separate from their litter. It is also necessary to pay close attention to rabbit behavior. For example, a rabbit in heat will be more restless and exhibit behaviors such as rubbing its chin on its feed. Other physiological signs like a red, swollen vulva will also become apparent. Once your rabbits are sexually mature, they should be separated into breeding pairs and kept away from the litter they came from. Parents should not be allowed to mate with children, and brothers and sisters should also not be allowed to mate.

Sustaining a Healthy Population Size

Maintaining a healthy population size means that you should be aware of the maximum number of rabbits that can fit in the space that you have. You also need to be mindful of the doe-to-buck ratio and keep it balanced. For example, there should be about two bucks for about 20 does. When you are mating your bucks and doing so, you must be aware that the bucks can become territorial. Therefore, your does need to be moved into your buck's territory instead of moving your buck into a doe's territory where other bucks are present. A doe and her litter need at least six square feet of space to be healthy. Therefore, the surface area you have in which to raise your rabbits needs to be measured so you can calculate how many rabbits are your maximum capacity. Rabbits vary in size, so the space they need will change. An excellent way to measure if you have enough space for humane breeding is that each individual rabbit requires a section that is about five times bigger than its body.

Your breeding practices will also determine how healthy your population is. Genetic defects can leave you with a decreasing population. Therefore, when you mate any of your rabbits, you need to check if they are healthy and whether there is anything abnormal in their development. You want to use your strongest and healthiest bucks and does for mating. Furthermore, you should regularly change your bucks. Over time, your does will also begin producing smaller litters. Keeping track of your litter sizes is another way of making sure that a strong population is maintained. Once your doe's litters begin to decrease significantly, it may be time to bring in younger does for breeding.

Reproductive Cycle

Ethical practices around the reproduction cycle are mainly concerned with the health of your breeding rabbits, as well as the conditions where breeding takes place. Rabbits are viable for breeding for about three years. It is suggested that you should swap your bucks at least once a year for optimal breeding. The span of a doe's pregnancy is about one month, depending on the breed. Once the kits are born, they nurse for approximately eight weeks. To ensure that your kits grow up healthily, they must breastfeed for the entire eight-week period. Kits can eat solid food at about two weeks old. However, this does not mean they are ready to stop drinking their mother's milk. A big part of ethical breeding is ensuring that you consider the health of your rabbits. Therefore, extensive care is needed for kits, as well as pregnant ones.

Once your doe has given birth and breastfed her kits, you need to make sure that she is healthy before breeding her again. Pregnancy takes a lot out of a doe. You can check her energy levels, as well as her body, for any injuries before breeding her again. You need to wait at least 35 days before rebreeding your doe after the litter has been weaned. This will protect the health of your doe, as well as the health of any future litter. Sometimes, a pseudo-pregnancy can occur, which lasts for about 17 days. This can happen if a doe mates with a sterile male or from other physical stimulation. Therefore, it's essential to monitor a pregnant doe if you want to maintain a healthy population.

Appropriate Breeding Ages

Rabbits are ready to breed from anywhere between four to seven months, depending on their size and their breed. Smaller breeds tend to mature faster. For almost all breeds, bucks mature much slower than they do. When your rabbits are ready for breeding, it is crucial to track their mating patterns to ensure that both the does and bucks remain healthy. A buck should be allowed to mate at least every three to four days. Healthy bucks can continue mating for about two to three years, but for a strong lineage, it is suggested that your mating bucks get swapped at least once a year. Unlike many other mammal species, rabbits do not ovulate according to a set schedule. Does only ovulate when there is sexual stimulation.

Commercial rabbit farms produce about five to six liters a year. For ethical breeding practices, your breeding should not be strictly driven by maximizing litters. Your dose should be monitored to make sure that they are healthy and strong enough to produce more litters. You do not want to put too much strain on your animals if your aim is to increase their quality of life. Although your rabbits are essentially a commodity, for your breeding practices to hold a high ethical standard, you must consider the rabbits' quality of life. Therefore, your rabbits should only be bred when they are in optimal health, and you should prevent overbreeding.

Caring for Pregnant Does

Pregnancy is a vulnerable time for most mammals. So, a higher degree of care is needed when a doe is pregnant. There are various ways to tell when a doe is pregnant. One of the more obvious ways to detect it is by checking the size of the abdomen. The body weight will also significantly increase. Another way of checking is by placing a buck near her. Bucks will not mate with those who are already pregnant. When a rabbit gives birth, it is known as kindling. Before kindling occurs, it is important to build a nest.

Sometimes, it can be difficult for a doe to conceive. This indicates that either the buck or the doe is unhealthy, and one of the main causes of a doe being unable to conceive is if she is overweight. Since it is ethical to maintain optimal health for your rabbits, you must ensure that your rabbits are at a healthy size. Overweight bucks also tend to lose libido and are lazy. Old age, disease, and injuries can prevent rabbits from breeding effectively. So, it makes sense that you keep your rabbits in tip-top condition.

Caring for Newborn Kits

Rabbits care for their young very well. Your duty as a breeder is to make sure that the conditions for caring are set up appropriately. Rabbits use their fur to make nests for their kits. You can add sawdust to the enclosure for the doe to use to construct a nest as well. Sometimes, rabbits can get orphaned. Maybe the doe got injured or died during the kindling process. If this happens, you will have to bottle-feed the baby rabbits using a special formula. A great substitute for rabbit milk is kitten milk. You will also need to construct the nest yourself.

Researching what is best for your rabbits is not a once-off activity. As you continue on your breeding journey, you will need to do continuous research. Keeping yourself up to date with the latest information is the cornerstone of ethical breeding practices. There are always new information and best practice updates available. For the most part, caring for newborn kits simply means checking on them daily and making sure that the mother is healthy because they are natural nurturers. If you notice any concerning changes or behavior that is outside of the norm, conducting research is beneficial. Through keen observation and staying updated with scientific developments, you can maintain a healthy herd.

Health Considerations

You choose the rabbits you raise; they don't choose you. Therefore, any health concerns land firmly on your shoulders. Rabbit meat is cheap, and rabbits are relatively easy to raise. However, that does not mean that they are able to remain resilient when their health is neglected. Rabbit health is primarily based on three factors: housing, nutrition, and maintenance. By grooming your animals, feeding them well, and creating a shelter that caters to their basic needs, you can ethically raise rabbits that will provide you with ample meat for profit or consumption. There are differences between raising rabbits for pets and raising rabbits for meat. Meat rabbits are bigger than rabbits bred for pets. Furthermore, there are differences in the care they need.

Meat rabbits will inevitably be slaughtered. However, just because they will be killed for their meat doesn't mean their lives must be filled with suffering. The better rabbits are taken care of, the healthier they will be. Healthy rabbits produce higher-quality meat that will taste better and can be sold at a premium. So, keeping your rabbits healthy is beneficial all around because they will have a higher quality of life, and it is beneficial to you because you will have a higher-quality end product. Ethical health considerations may take more time and effort, but it is worth it for the animal welfare, as well as the premium meat that you will enjoy.

Nutrition

Commercially available rabbit feed in the form of pellets has all the necessary nutrients to maintain a healthy diet. Using commercial rabbit feed is more advisable than trying to mix your own feed because it has been scientifically formulated. Treating your rabbits with some leafy greens or veggies occasionally is okay, but commercial pellets are sufficient for all their dietary needs. Additionally, water must always be readily available for the rabbit. Since rabbits are susceptible to extreme temperatures, a lot more water is needed in the summer to help them cool down and control their body temperature.

Commercially available rabbit feed in the form of pellets has all the necessary nutrients to maintain a healthy diet.
4028mdk09, CC BY-SA 3.0 <https://creativecommons.org/licenses/by-sa/3.0>, via Wikimedia Commons: https://commons.wikimedia.org/wiki/File:Jungtiere_Kleinsilber_Kaninchen.JPG

Housing

Rabbits are sensitive to extreme temperature changes. Therefore, housing is one of the main factors that contribute to healthy rabbits. Increases in cold or heat can cause rabbits to die or become infertile. Enclosures must also be built in a way that blocks the wind. The combination of wind and cold is devastating for rabbits, and you could end up losing your entire herd. Ethically, as a breeder, you must create a comfortable space for your rabbits. A well-kept rabbit enclosure allows your rabbits to be comfortable, safe, and free from injuries and diseases.

In addition to maintaining warmth for your rabbits, fencing is a central part of rabbit housing. Many people make the mistake of thinking that fencing is primarily meant to keep rabbits in, so they buy the cheapest fencing from their local garden store. This approach is completely wrong. Any fencing that is installed must be predator-proof.

Rabbits are vulnerable to all kinds of predators, including foxes, coyotes, dogs, cats, and various birds of prey. Your fencing should be geared around keeping predators out more than keeping your rabbits in. It is not ethical to put your rabbits in harm's way because it's your job to ensure their safety. Healthy and safe living conditions for your rabbits are the crucial pillars for breeding rabbits ethically.

Medical Care

Preventing illness, disease, and injury is centered upon creating sanitary conditions for your rabbits. Your rabbits and the environment that they live in must be kept clean at all times. Rabbits have sensitive ears and nails, which means that you need to check on them regularly. Furthermore, you need to keep your rabbit's nails short. When clipping their rabbit, make sure you don't nick any arteries by cutting below the quick. This is excruciatingly painful for the rabbit, and it could become infected. Getting regular checkups from a vet is also advisable. If one of your rabbits gets a disease or illness, they can likely spread it to all your rabbits. So regular vet checkups are advisable so that you can pick up on any illnesses early on.

Chapter 7: The Life Cycle of a Rabbit

Understanding the life cycle of rabbits is essential to maintain a high-functioning breeding operation. At different times of their lives, rabbits display certain behaviors and needs. To provide their rabbits with the best care to produce quality meat, a breeder needs to be acutely aware of the stages of rabbit life. There are four main stages of developmental evolution: the kit or newborn, the juvenile, the adult, and the senior phase. Grasping the intricacies of each phase of the rabbit life cycle helps a breeder make informed decisions that will benefit their herd and, ultimately, their meat.

At each stage of their life cycle, rabbits will have specific needs. A breeder must effectively meet these needs, and breeding healthy rabbits requires an in-depth understanding of their biological functioning. Age plays a big role in the biology of a rabbit. Considering that a breeder interacts with rabbits from birth to adulthood, understanding what rabbit behavior and functioning is normal at each age can help them determine how fit their herd is. A healthy herd leads to better meat or more profit if a breeder aims to sell their rabbits.

Kit

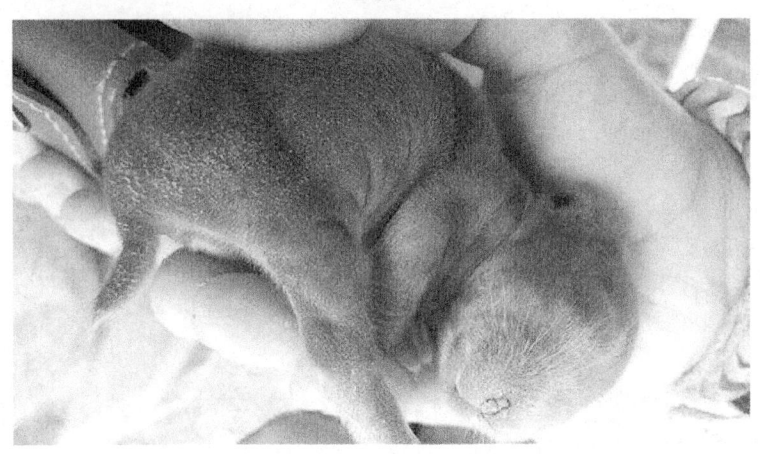

The kit stage is when a rabbit is at its most vulnerable.
Shuluh Shasa Nadita, CC BY-SA 4.0 <https://creativecommons.org/licenses/by-sa/4.0>, via Wikimedia Commons: https://commons.wikimedia.org/wiki/File:Newborn_bunny.jpg

The kit stage is when a rabbit is at its most vulnerable. Kits require specialized care from both the breeder and its mother. Knowing exactly what a healthy kit looks like and how it behaves leads to more informed breeding decisions and interventions that might be needed at this early stage of life. When you are breeding rabbits, there will be a constant cycle of newborns that need your attention. The nutritional requirements of the kits, coupled with housing and medical care, create a matrix of focused attention needed to assure the best possible outcomes for your herd.

The kit or newborn stage is from birth to about three months old. Typically, at this part of the rabbit life cycle, a high level of care is needed from the dam or the mother rabbit. The fragile babies clumsily scurry around, clueless about the dangers of the pen. The sire, or the father rabbit, does not play a pivotal role at this level of development. Kits are born with their eyes closed and have no fur at the early stages. Their body heat cannot be maintained, so they are kept warm by the dam and a nest. Breeders need to provide sawdust, which the dam will use together with her fur to keep her young warm.

For the first two weeks, kits acquire nutrition that comes predominantly from milk. After the first two weeks, they will start eating pellets. However, they are not ready to be weaned off milk until about eight weeks. A newborn rabbit will be two to three inches long and will weigh 30 to 40 grams. These gorgeous babies feed once or twice a day for five to ten minutes per feeding. The nutrient-rich milk provides the appropriate sustenance for a young rabbit on one feeding every 24 hours. Kits will usually feed early in the morning, between midnight and 5 a.m.

Since newborn kits are so vulnerable, it is essential to check on them consistently, especially immediately after the birth. The dam sometimes leaves the nest unattended, which is a perfect time to check on the little ones. Unfortunately, it is common for the babies to die early on, so this is an opportunity to remove any dead kits. Handling the nest must be done with extreme caution so as not to disturb them too much. A dam will eat the placenta after birth, but sometimes, the cleanup can be a bit sloppy. Assist mom by removing any leftover placenta when you are checking on the newborn kits. Once you have made sure that all the kits are healthy and alive, you can check if they have been fed. A dam can have a problem feeding kits, so they will have to be bottle-fed. If they have been fed, their bellies will be round and protruding. If you disturb the nest, it is essential to restore it to the way you found it.

The young ones will finally be ready to leave the nest at about two to three weeks old. At this point, they begin eating pellets and will not be exclusively reliant on their mother's milk. After six to eight weeks, your kits can be removed from their mother and weaned off milk. The babies will now be much more independent and will not require the focused care they once did. However, rabbits are still considered kits up until about three months old. Although these young ones are not as vulnerable as they once were, it is still necessary to check on them regularly to monitor their health.

For the health of the dam, weaning should be done gradually. When kits are abruptly removed, this can cause the development of mastitis. Mastitis is a disease that results in the swelling of the breast of the rabbit. This swelling can lead to infection, resulting in the mother's premature death. Although the sire or the father is not as integral to the development of the kits as the mother is, it can also help to keep the sire around for the first few weeks to prevent stressing out the babies or the dam. The father acts as a stable foundational support for your breeding family.

Juvenile

Once a kit has grown significantly, it transitions into a juvenile. This stage can be likened to a human teenager. The juvenile phase is the bridging period between being a kit and reaching adulthood. This stage starts from about three months old up until about a year. Some of the bigger breeds reach adulthood earlier, at about nine months old. Feed becomes more important as hungry teenagers consistently munch to support their rapid growth. Alfalfa hay and quality pellets are a good feed to use at this stage of development because it is so calcium-rich and will help adolescents develop strong bones and muscles. It's also high in fiber, which aids in digestion.

At this stage, feeding should be constant, with a consistent flow of pellets readily available. The explosive growth of juvenile rabbits requires nutrients, so they feed often. The most important consideration for raising rabbits at this age is making sure that they have ample food and water readily available. A good balance must be maintained by providing free-flowing feed and, at the same time, preventing overeating because these young ones can be gluttonous. The chubby juveniles can sometimes be prone to gorging themselves. You need to develop a feeding protocol that makes sure that the rabbits are well-fed while not allowing them to overindulge.

Juvenile rabbits are very active, so they require a lot of space to run around. Creating an enclosure with different levels that the rabbits can hop around on will help manage their high energy levels. As they transition to adulthood, they experience hormonal changes, and you'll notice they become much more aggressive. Young rabbits may exhibit unruly behavior like damaging their enclosure, biting, or spraying urine everywhere. Therefore, these unpredictable guys must be monitored to make sure that they do not hurt themselves or other rabbits. Biting can cause injury or infection, so check to make sure that your rabbits do not have wounds caused by your hormonal juveniles.

Rabbits that are kept for pets are often neutered when they reach this juvenile phase because of their hyperactive behavior, like biting, digging, and moving around frantically. This behavior must be tolerated by a breeder because you need these misbehaved rabbits to reproduce. So, any shelter that is built for these teenage nightmares should take into account the rebellious and active nature of rabbits at this age. You can also expect more fights from adolescent rabbits due to their hormonal changes. At this developmental phase, the mischievous rabbits tend to express aggressive territorial behavior. They will lunge at you when you enter their space, so you have to be careful when handling them.

The curiosity, aggression, and hyperactivity that grow in the juvenile phase of rabbit development will subside with age. These behaviors are influenced by the raging hormones of puberty. Juvenile rabbits are moving closer to sexual maturity. In male rabbits, you may find that they begin mounting objects or other rabbits. This is a normal sign of development for a young rabbit. After the age of one, they will step into adulthood, so they will calm down significantly from the peak of the rambunctious transitionary stage of their life cycle. Male rabbits are more aggressive than females and are the ones who express the most territorial and aggressive behaviors.

Hormonal shifts also play a role in female rabbits. As juveniles, female rabbits may begin nesting. Adult rabbits only nest when they are ready to give birth. When young females begin nesting, it is a sign of false pregnancy. The rabbit is getting ready to start mating. Rabbits do not ovulate when there are no males around. The hormonal changes are preparing the animal for sexual maturity and the biological demands that come with it.

The primary concerns when rabbits are at the juvenile phase are preventing fighting and maintaining their living spaces. Juvenile rabbits are prone to damaging property by chewing and digging. Furthermore, your animals may need vaccinations at this early stage to prevent the spread of various diseases. It is advised to either have a vet do a home visit or to take your herd to the vet so that a medical professional can make sure that they have no injuries or life-threatening diseases. If a small number of rabbits get sick in your herd, they can likely spread the illnesses. Therefore, getting all the necessary vaccinations early on is highly recommended.

Adult

When rabbits reach adulthood, it means that they will no longer increase in size.
https://www.pexels.com/photo/white-rabbit-wearing-yellow-eyeglasses-4588065/

Depending on the breed of the rabbit, adulthood is reached between the ages of nine months to 18 months. This is when your rabbits are fully grown and ready to begin breeding. The life span of a rabbit is from about three to nine years; therefore, most of a rabbit's life will be spent as an adult. Meat rabbits called fryers are usually slaughtered before they reach adulthood at the ages of about three to six months old. The adult rabbits that you have will predominately be used as breeders. Some people deem it more ethical to wait for a rabbit to reach adulthood before they get slaughtered or sold for meat.

When rabbits reach adulthood, it means that they will no longer increase in size. Now that you know your rabbits will not get any bigger, you can calculate the dimensions of your enclosures according to the mass of the mature rabbits. In adulthood, rabbits are not as aggressive as they were when they were juveniles, but they can still display some of the same behaviors, like being territorial. Considering that your adult rabbits are your breeding rabbits, they are essentially the center of your operation. Therefore, specialized care is needed to make sure that they are able to breed well.

Food for adult rabbits needs to be strictly controlled. Unlike juvenile rabbits that are still growing, adult rabbits have reached their full size, so their food supply will be steadier. High-quality pellets and hay are all that is needed to feed adult rabbits. In addition to the food, they should also have readily available water. Rabbits tend to become lazier when they get older, so there is a risk of obesity, which can affect their fertility. Since your adults are your breeders, it is essential to make sure that they maintain a healthy, productive weight to maximize litter sizes.

The rabbits maturing to adulthood are carefully selected by breeders. Your adult rabbits should have all the most desirable traits that you intend to pass on to future generations. All the rabbits that you keep until adulthood should be the strongest, fittest, and healthiest out of your litter. These are the rabbits that are not prone to injury or illness and can produce the most meat. Therefore, you must be carefully selective of the rabbits that you sell or consume and the ones that you keep for breeding. Usually, there are about two bucks to twenty does, so you will have to select your rabbits according to gender as well.

Considering that adult rabbits can become lazy, which could lead to them gaining weight, your enclosure needs to be structured in a way that encourages activity. Installing some moving parts into your housing can be beneficial because a repetitive environment can cause boredom amongst adult rabbits, which will fuel their lazy proclivities. When you have sections that can move around and allow you to rearrange your enclosures, it will prevent laziness and keep your rabbits active. Movable parts that you can install in your enclosures can be simple, like steps, boxes, or tunnels for your rabbits to play in. Nutrition and exercise are the two most important factors when caring for adult rabbits.

Sick rabbits, or animals with genetic deficiencies, will usually be culled before they reach adulthood. Some rabbits with deformities can still be sold for meat. Considering that your genetic selection will make sure that you have a powerful bloodline of adult-breeding rabbits, it will not be difficult to care for them at this age. Besides cleaning, feeding, and checking for injuries, adult rabbits are relatively low maintenance when compared to newborns, juveniles, and even senior rabbits.

In adulthood, your rabbits will become a lot less active after about three years. This is when aging starts to kick in, and they are transitioning to the senior geriatric phase. This is why it is suggested that your breeding males should be swapped at least once every year. Furthermore, as females age, their litters will also become smaller.

Therefore, when your rabbits are adults, you need to monitor how active they are, as well as the size of the litter, so that you can make profitable breeding decisions. Adult rabbits are low maintenance, but the care you give them will determine how long they can effectively breed.

Senior

For a breeder, it is unlikely that their rabbits will reach the senior phase of life. Rabbits are considered geriatric between the ages of five and seven. At this age, your rabbits are close to natural death. Therefore, as a breeder, your animals will get sold off before they reach this late stage of life. However, if you keep some rabbits until they are geriatric, they do have specific needs that are unique to elders. Thus, you will need to adjust your care once any of your rabbits have reached this stage of their life. The needs of senior rabbits are excessive, and it could be compared to the level of care needed at the newborn level.

A senior rabbit is going to be a pet. No profitable benefits are attached to caring for a senior. When a rabbit becomes geriatric, they are no longer able to breed. They can be sold for slaughter, but meat rabbits are sold long before they reach this stage. Old rabbits become even less active than adult rabbits as their health begins to slowly deteriorate. Many people keep older rabbits for pets because of how calm they are. Senior rabbits are highly unlikely to lunge at you like juveniles and do not exhibit the same kinds of territorial behavior. Moreover, they are not damaging to enclosures because they are so inactive. Senior rabbits eat less than adult rabbits because their appetites decrease with age.

Geriatric rabbits are difficult to take care of because they experience a number of diseases and ailments. These elders can have ear issues, renal failure, arthritis, eye problems, and dental diseases. Female rabbits that have been breeders and have not been spayed can also develop uterine tumors. Therefore, visits from the vet will be common for senior rabbits. Older rabbits are fragile and require constant care. Since they no longer breed, keeping a senior rabbit is a liability to a breeding operation, so if you are going to keep rabbits until they get this old, you need to be aware that they will cost you money.

Older rabbits tend to get scared easily, so you'll need to keep them in a stress-free environment. Geriatric rabbits can suffer from heart failure when they are frightened or startled. Rabbits are social creatures, but keeping a senior rabbit around many adults and juvenile rabbits could be fatal because of how high the stimuli in that kind of environment are. Any decisions you make about the living conditions of your senior rabbits will have to include considerations for the fragile heart conditions of the vulnerable older rabbits.

To extend the life of a senior rabbit, the environment it lives in must be comfortable. Senior rabbits also need some exercise, but, for the most part, they will be lazy. A soft environment with carpets and cushioning is ideal to prevent injury, as well as provide some comfort in movement because their joints may be sensitive. The loss of mobility caused by their deteriorating joints means that they will not be able to groom themselves well. Therefore, if you are keeping older rabbits, you will need to groom them and give them dry baths.

Senior rabbits rapidly lose weight as they begin losing their appetite. It becomes essential to feed senior rabbits with high-quality and nutrient-dense food. The loss of appetite that older rabbits experience contributes to them rapidly losing weight. Caring for senior rabbits means that you consistently have to check their physical condition because changes can happen quickly. While you are grooming a senior rabbit, you can check for any lumps or injuries because these can be indicators of a number of diseases. Furthermore, senior rabbits need their nails trimmed more often because their inactivity allows for their nails to grow a lot longer. The combination of grooming, nutrition, medical care, and checking for injuries is the reason senior rabbits require more attention than younger rabbits.

Chapter 8: Compassionate Harvesting

Breeding rabbits for meat means that, at some point, your animals will be harvested. Furthermore, some of your herd will need to be euthanized for various other reasons like population control, genetic defects, or diseases. Killing does not need to be cruel. Breeding rabbits for meat makes killing unavoidable, but there are ways to ethically approach the process of harvesting. Educating yourself on the most humane practices in the industry can help you set up a more compassionate operation. Your animals provide you with meat or income at the expense of their lives. Therefore, you are responsible for making sure that your rabbits have the best experience possible before they are forced to make the ultimate sacrifice.

Any unnecessary distress needs to be eradicated from your breeding operations. This suffering includes injuries and prolonged deaths. Slaughtering and euthanasia should be done in the most painless ways possible. The ASPCA says that for a slaughter to be considered humane, the death has to be painless, or the animal's senses must be numbed (Browning and Veit, 2020). Furthermore, the ASPCA also advocates for instantaneous death that is free from agony (Browning and Veit, 2020). Numerous measures can be taken to ensure that your breeding setup meets or exceeds these standards.

In addition to painless slaughter and euthanasia, the conditions in which they live leading up to the killing of your animals must also be taken into account. Your rabbits need to be kept calm and must be put at ease before slaughter. Therefore, using humane handling techniques and crafting a slaughtering environment that is geared toward compassionate killing is crucial to running an ethical backyard farm. Many people are driven to backyard farming to produce their own meat because of some of the atrocious practices that are widespread throughout factory farming. Farming meat rabbits sustainably and conscientiously requires you to embrace reform in the farming industry by implementing change on a small scale.

Educating yourself on the most humane practices in the industry can help you set up a more compassionate operation.
https://www.pexels.com/photo/macro-shot-of-heart-shaped-cut-out-1820511/

Compassionate harvesting does not begin at the butchering table but encompasses all your breeding practices. The reduction of stress before butchering is incumbent on the conditions that your animals live under. Therefore, a safe, sanitary, and spacious environment is the initial step to compassionate harvesting. Compassion implies a high level of care for your rabbits. Thus, the killing process must be conducted with the utmost respect for the animals. There is no truly kind way to kill an animal. However, you can get as close as possible to humane by being principle-driven over commodity-driven. Rabbits are mammals, so people can somewhat relate to their ability to experience pain and suffer as sentient creatures. The connection you have with your rabbits as living beings providing you with sustenance in the form of meat or profit must be shown through caring practices.

Inhumane slaughter and euthanasia, where animals experience excessive pain and suffering, can be reduced by effective training and information. Human error is largely to blame for inducing animal suffering through cruel slaughter practices and mistakes being made in the stunning process. Getting professional assistance and working toward being competent in slaughtering and euthanasia can prevent animal anguish. The rabbit industry is largely unregulated due to the meat not being as popular as beef, chicken, or mutton. However, this weak regulation should not encourage cruel practices. The lack of proper regulation in the rabbit breeding industry should be an encouragement to hold yourself to an even higher standard as you are in the position to contribute to industry standards as a way-paver.

Humane Methods of Euthanasia

Rabbits have complex physical, social, and psychological needs. When killing rabbits, you should consider all these multiple variables if you aim to make the breeding operation ethical and humane. Slaughtering animals for meat may seem like the primary concern of killing animals. However, breeding will sometimes require killing for other reasons. Profitability is a goal for many breeding operations. There are times when it can be difficult to sell rabbits, which could result in the need for some animals to be culled.

Overpopulation in a small space is a form of animal cruelty. Therefore, euthanizing rabbits can be perceived as a way to show mercy to your herd. The process of selectively reducing your population is called culling. There are two types of culling when it comes to meat rabbit breeding, namely hard culling and soft culling. Soft culling is when your rabbit population is not killed. A soft cull includes selling rabbits for pets or halting breeding activities. Hard culling refers to euthanizing your rabbits. When you cull your animals, you will choose weaker rabbits that have defects or seem to be more prone to sickness.

There are a variety of ways that breeders select which rabbits will be culled. One of them is the elimination of those without mothering instincts. Sometimes, a dam will abandon its motherly duties like cleaning its kits after birth. Dams without mothering instincts often trample or eat their young. Many breeders go by the three-strike rule, meaning that if a rabbit displays an inability to mother three times, it puts the animal in a position for culling. People get attached to the animals they raise, so this can be a difficult decision to make. A meat rabbit breeding operation is dissimilar from a sanctuary or pet breeding operation in the sense that culling can become part of the business model.

The most common way to euthanize a rabbit is by putting it under anesthesia, whether this is through inhalation or by injection, and then proceeding to decapitate the rabbit. The rabbit will not experience any pain or suffering when this method is used. Another way that rabbits are euthanized is by lethal injection into their main vein. This method also painlessly kills the rabbit. These methods of euthanasia are conducted by a qualified vet. It can be dangerous to attempt these methods yourself as an unqualified person because a mistake can result in the prolonged suffering of a rabbit.

Using a vet to euthanize your rabbits can be expensive, so some breeders will opt to euthanize their own animals. For young kits, it is easier to kill them because they are smaller and more fragile. A swift motion with a sharp knife to decapitate a rabbit, starting from the spine downwards, can end the animal's life relatively painlessly. With older rabbits, using a knife is not as humane because the spine is strong enough to prevent a clean cut. With younger rabbits, you can also use the bowl method for euthanasia. The bowl method is sometimes used by pet shops for culling mice and rabbits, which are fed to snakes. Place the bowl on the back of a kit's neck. Push down hard on the bowl while you pull the kit's hind legs. This method dislocates the kit's cervical, which produces an instant kill. Some people use carbon dioxide chambers to euthanize rabbits, but this method is undesirable because death is slow.

Compassionate Butchering

Your rabbits are bred for meat, so they will need to be slaughtered at some point. You can outsource the slaughter to a more skilled individual, or you can learn how to slaughter the rabbits yourself. When you outsource your butchering, there are two main factors to consider.

Firstly, the person that you appoint must be knowledgeable about how to kill your animals without causing suffering. Secondly, some cost considerations need to be made because you will most likely be paying for the service. Therefore, if you are selling your rabbit meat, the slaughtering process will need to be accounted for in your sales price.

You have the option of selling your live rabbits to butchers. If you take this option, you need to also do some research about the condition of the slaughterhouse. It is advisable to do a tour of the place to see if it meets your ethical standards. The rabbits will need to be stunned before killing, or if they are not numbed, a sharp knife should be used to ensure that the kill is instant. Moreover, it is essential to check the cleanliness of the abattoir because unsanitary conditions can cause further suffering of animals and contamination of the meat, which causes discomfort for the people who consume the product.

Homestead breeders also use a variety of other methods for slaughtering rabbits. One of the more prominent methods is using a blunt object to hit the rabbit hard on the head. Hitting the rabbit on the head causes the animal to be knocked unconscious. After the rabbit is knocked out, a sharp knife will be used to cut one of the main arteries in its throat and bleed it out. The problem with using blunt force to knock a rabbit out is that if you hit the rabbit incorrectly and the strike is not precise, the rabbit will not be knocked unconscious with the first hit. Suppose the rabbit is not made unconscious after the first blow. In that case, multiple blows will be needed to knock the animal out, which causes pain and trauma before slaughtering.

The most humane method of slaughtering is dislocating the cervical (neck). To do this, use a sharp knife to cut the spine while the rabbit's hind legs are pulled. This method is more difficult to use on older rabbits because their spines are stronger, so the knife has to be extremely sharp, and a lot of pressure must be applied. Furthermore, the method takes skill, so it is unlikely that a beginner will be able to painlessly slaughter a rabbit using this method. If you are going to dislocate the cervical of your animals to ensure an instant kill, it's best to do it with the guidance of a professional familiar with the process. A few lessons before attempting the method on your own could be highly beneficial to maintain a great ethical standard.

Rabbits are social animals, so removing those you are slaughtering away from the group is recommended to reduce stress. Furthermore, considering that rabbits can be prone to heart problems, as well as being sensitive to extreme temperatures, it is essential to make sure that the environment for slaughter is calm and temperature controlled. Your rabbit should be restrained properly to prevent movement, which could lead to inaccurate slicing that harms the animal. The conditions that you slaughter the animal under are just as important as your slaughtering technique.

Minimizing Stress

The stress of rabbits is often related to the environment in which they are raised. Many breeding operations use small cages crammed into limited space so that profit can be maximized. This creates a stressful environment for rabbits where they can exhibit antisocial behaviors like hurting themselves and others. Rabbits are socially and psychologically complex and need to be placed in a humane environment to allow them to thrive. One of the more humane ways of raising rabbits is called cuniculture farming. In this form of rabbit breeding, the animals are given space to graze naturally as if they were in the wild. This can be more costly because more space is required for fewer rabbits to be produced. However, this is one of the best ways to give a rabbit a stress-free life before it gets slaughtered.

The enclosure that your rabbits are kept in should be clean, spacious, and safe. Having an unsafe environment could result in your rabbits picking up illnesses or getting injured. These diseases and injuries that occur in inhumane living environments add to the suffering that rabbits experience before getting slaughtered. An acceptable enclosure should have enough space for rabbits to move around in and should preferably have multiple levels for rabbits to jump on and crawl through. The fencing should also be installed in a way that has no protruding parts that can cut or injure the rabbits. Humane slaughtering conditions require sufficient housing.

Since rabbits are social, interacting with them often can build enough trust that your presence calms them down. Grooming your rabbits and interacting with them during feeding can help them to bond with you. The connection you establish by building a relationship with your rabbits means they will be calmer when they are taken to get slaughtered. As a compassionate breeder, you must reduce rabbit suffering all the way until they end up on the chopping block.

The psychological state is just as important as the physical condition. Rabbits that are in a poor mental condition could be further impacted physically. The behavior of your rabbits will show you the mental condition that they are in. For example, distressed animals will have altered toilet habits and will chew on their cage or even repeatedly circle their enclosure. When herd members exhibit these kinds of behaviors, it may be a sign to adjust the way that you are caring for the animals. Maximizing the comfort of your rabbits will impact your final product by giving you high-quality meat because the mental and physical health of the rabbits contribute to how they develop.

If the environment that you are breeding your rabbits in is stress-free for the animals, it can also support your breeding operation by providing bigger litter. Fertility can be linked to psychology because distressed rabbits can have heart issues that impact their mating. It is not only the rabbits that benefit from consideration for their mental state, but you as well because your breeding rabbits will produce bigger and healthier litters if their mental state is at ease. By keeping a calm and healthy environment, you can produce meat to be sold at a premium. Having premium meat available can be profitable, especially considering that the rabbit meat market is smaller than other domestic animals like chickens, sheep, or cows.

Ethical Animal Treatment

The need to butcher and euthanize animals painlessly to prevent suffering is built upon the principle of ethical animal treatment. Many organizations in the farming industry create regulations that govern how farmers can treat animals. Various activist groups challenge some of these regulations because they are conscious of the well-being of animals. Since the rabbit breeding industry is not as strictly governed as other meat farms, the ethical treatment of your animals is placed on the shoulders of individual breeders. The initiative to take the precautions to build a humane backyard farm is largely driven by your own ethical approach.

Research into the psychology and complex societal structures of rabbits can be used as a guiding tool to create an ethical farm where compassionate harvesting is promoted. The aim is to understand rabbits enough to create an environment where they are able to physically and psychologically flourish. Although your rabbits are, in essence, a commodity, that does not mean that their well-being should be ignored.

It is in your best interest to treat your rabbits ethically because healthier rabbits will produce better meat.

Breeders have an obligation to be caring – and various regulatory bodies govern how you can treat your animals. Aligning with the law is only the first step in being ethical. Compassionate harvesting implies that there is an emotional connection to the animals that you breed. Bigger scale operations tend to create a coldness due to the methodological factory approach they have to breeding and slaughtering rabbits. As a backyard breeder, you are in the unique position of using the small-scale nature of your breeding operation to create a more personalized style of care. Unlike your factory farm counterparts, you can take the time to care for each individual rabbit to make sure that they are living their best lives.

The emotional connection you establish with your animals, coupled with being well-researched, informed, and educated, builds the foundation for compassionate harvesting. The difference between compassion and indifference is the ability to feel the pain of others. Therefore, establishing a strong bond backed by scientifically informed knowledge can put you in a position to understand the pain and desires of your rabbits. From the ledge of understanding, you can overlook a backyard breeding setup that is humane, profitable, and functional.

Chapter 9: Using Rabbit By-Products

Rabbit meat, when compared to other meat, is a rich source of protein and is healthier than most. Beyond their delicious meat, rabbits are also cultivated for their by-products – which are profitable and can give you a regular income stream. Therefore, this chapter teaches the ethical and efficient use of all parts of the rabbit and the beneficial use of rabbit manure for gardening. Additionally, you will be guided on using other by-products like bone and organs.

Ethical and Efficient Use of All Parts of the Rabbit

Head: In some countries, the head and brain of rabbits are eaten. Recipes like spicy Sichuan Rabbit Head and Rabbit Head Paste are examples of recipes using the head and brain. Traditionally, rabbit head is used in stews and for stock. They are also used to feed dogs, pigs, and chickens. Rabbit heads are crushed for chicken feeds with the blood, meat, and bone regarded as a good choice for feeding laying hens.

The rabbit brain is used in the pelt tanning process. It is believed that the brain size of every animal is sufficient to tan the pelt of that animal. Furthermore, the brain is a rich source of omega-3 fatty acids.

Ear: Rabbits ears are dehydrated and used as a dog treats. They can also be deep-fried and eaten with apricot ginger chutney sauce.

Pelts: It produces blankets, headgear, coats, hats, and other clothing for keeping warm. They can be added to clothing as trimming.

Feet: You can turn rabbit feet into a lucky charm by drying them and adding some cool decorative items. You can do this by adding in a small jar 70% isopropyl rubbing alcohol. Soak the feet completely in the alcohol solution for two days, creating a lock in the fur. This alcohol will dehydrate the cells, killing fungus and bacteria. Rinse with clean water after two days. Mix some borax with water to a ratio of 15 to 1. You can use hot water to get the borax to dissolve quickly. With its antibacterial and antifungal properties, the borax would dehydrate the tissue and skin to preserve the foot. Ensure you completely soak the feet into the mixture, leaving it for a day. After one day, take the foot out of the mixture to dry in the sun. Brush clean and add beads or any decorations you choose. Rabbit feet can also be frozen or dried for dog treats.

Tail: For centuries, the rabbit tail has been used in pollinating flowers. You achieve this by attaching the tail to a stick, rubbing it over the male and female flowers, and transferring pollen. Furthermore, the tail is used for making keychains, zipper pulls, and dog treats.

Blood: The rabbit blood is used for making blood pudding and sausage. You can use the rabbit blood to make charcuterie and to thicken sauces. The blood can be mixed with sawdust to become soil additives or mixed with water and poured around your trees, shrubs, and bulbs to fertilize them.

Liver: The liver is used for making liver pate and contains enough iron. It's also used to feed dogs, chickens, or pigs either in its cooked or raw form.

Kidney: You can either eat the kidney alone because it's nutritious and tasty or make a rabbit pot pie, stuffing, and sausage out of it. Moreover, you can feed the kidney raw to your pets.

Heart: The rabbit heart can serve as a source of trace nutrients, B vitamins, and coenzyme Q-10. You can also feed them to your pets.

Lungs: You can feed the lungs to your pigs, chickens, or pets either in its raw or cooked form.

Stomach/Pancreas: They can be used as feed for your pigs or dogs.

Uterus/Testes: They are used as raw feed for dogs, chickens, or pigs.

Rabbit Manure: It is known as the world's best fertilizer for your farm or garden. It contains about 2% nitrogen, 1% potassium, and 1%

phosphorus.

Rabbit Urine: Rabbit urine mixes with water in a measurement of 10:1, fights off aphids, and fertilizes plants.

Bones: Rabbit bones make compost for bone meal fertilizer and prepare tasty rabbit stock.

Fat: It makes candles or soap and can be turned into lard or fed to chickens and pets.

Intestines: You can use the intestines to feed your pigs or dogs or dig a hole and bury them as they act as compost for your soil.

General Usage of All Parts

While the American market mainly uses the meat and dumps the rest, people from other parts of the world go beyond the meat and have found a use for every part of the rabbit. Rabbit meat gives you the protein needed for building your muscles, but the organs, when consumed, feed your organs. How? The rabbit bones, when blended, can act as a healing elixir that refreshes your digestive system. Bones and joints can be blended and made into bone broth.

The unsorted parts of rabbits that you don't need can either be fed to your pets or frozen and sold to others who may need them raw for their pets. Dogs and cats with allergies or health conditions should be fed rabbits. This is why the demand for rabbit ears, heads, organs, meat, and pelts is high.

Rabbit parts and their meat supply various prey-model raw and bone and raw food diets for dogs. With rabbits, nothing goes to waste.

The Rabbit's Fur

Every variety of rabbit has a textured fur that makes their wool or pelt production outstanding.
https://www.pexels.com/photo/selective-focus-photo-of-rabbit-2061754/

Have you ever admired your sweaters and appreciated the warm wool? Unknowingly to you, your gratitude should also be directed at the rabbit. For centuries, rabbit fur has been used for wool, which has blossomed into a significant fur trade. Every variety of rabbit has a textured fur that makes their wool or pelt production outstanding. There are different methods for removing fur from your rabbit. The first method is by burning the fur off the rabbit with fire. Another method is using hot water to peel the rabbit's fur. Thirdly, you can use the knife or without the knife technique.

Using a Knife

- Chop off the head of the rabbit or use a knife to slit its throat. It's one of the most humane ways of killing the rabbit. Another way to get the job done is by breaking the neck of the rabbit so as not to allow it to suffer.

- Just above the leg joints of the rabbit, cut a ring around each leg. At this time, the rabbit's legs should be strung on a rope. Do not cut deeply into the skin of the rabbit. Only cut enough to get to the hide.

- Make a single slice on each leg going up from the ring to the buttocks of the rabbit. It will make the skinning simpler at the end.

- Working your way from the ring cut you made earlier at the foot to the rabbit's buttocks or genital area, pull some of the hideaway. The hide should come off easily as you pull.

- Make your way through the bone of the tail by cutting through, ensuring that you do not puncture or sever the bladder in any way.

- Using both hands, begin pulling the hide from the rabbit's body. Like peeling a banana, the hide slips off easily at this point.

- Where the arm is, work your fingers into the sleeves of the hide, gently removing the arm from the hide. This could be challenging at first – but don't give up as it gets easier as you keep working your fingers through the sleeves.

- Continue pulling the hide from the upper torso to the head. Let the hide be resting at the base of the skull.

- Sever the head from the spine if you have not done it in the first steps. By this, the skin should be completely off the rabbit's remaining meat.
- Break the bones at the arm and leg with your hands, then completely remove the skin at the joint with your knife.
- Save the hides for tanning as needed while you dress and clean the meat.

Without Knife
- Place your hand around the rabbit's knee, pushing on the knee joint until it pops out of the hide, revealing the meat. You push the knee in one direction while pulling the hide in the opposite direction.
- With your finger, work your way around the leg till the hide is separated from the joint.
- While pulling the skin down, focus on pulling the knee joint upward till most skin is removed from one of the legs. This step can be likened to pulling down your pants (hide of the rabbit) and exposing your skin.
- Do the same thing to the other leg.
- Underneath the genitals, move your hands under the skin coming across the belly. Remove the skin from the belly by pulling it in.
- Put your hands on the buttocks area immediately above the tail and work under the skin to the back of the rabbit.
- Pull on the skin with both hands until it reaches the rabbit's arm.
- Break the skin between the head and the front arm with your fingers. Keep on pulling the sleeves of the hide upward, away from the arm meat.
- The spine should be cracked where it connects to the head.
- Keep the hide for tanning or other uses while you dress and clean the meat.

The Use of Hot Water
- Chop off the rabbit's head or strangle the neck to ease the pain of death.

- Put the rabbit in a bowl and pour boiled water into the bowl. Ensure that you pour the boiling water over the rabbit's body so the skin can come off easily.
- Allow the rabbit to remain in the hot water for 10 minutes. This ensures the rabbit's skin is properly soaked to help loosen it from the body.
- Start plucking off the fur from the rabbit's body. This should be easy as the water has been properly soaked into the fur.
- See to it that no fur remains and that the body is smooth by running your hands over the rabbit's body.

This technique is for those who are more interested in the meat aspect than the fur.

How to Clean Rabbit's Fur

When the skinning phase of the fur removal is over, wash off the hide in cold water to cool off right away. Do not worry about any tissue or fat left on it at this point. Your effort should be better spent on washing away the remaining blood on the skin because there is bound to be a permanent brown stain on the leather if the blood is not properly removed during this stage. If you are using soap or detergent, although not necessary, ensure traces of this cleanser are removed properly before proceeding to the next phase. Carefully extract the remaining water on the pelt once the rinsing is done.

Another way to clean the skin is with a washer. If there's a chance that bits of hair and fat will plug up the drain hose while using a washer, then avoid it and instead hand wash the pelt. It will allow you to examine the fur up close. When you're done thoroughly cleaning the hide, preserve it by drying on a stretcher, salting, drying, or freezing.

Uses of Rabbit Fur

Here are some of the uses of rabbit fur:
- Clothing
- Bedding
- Stuffing toy dolls
- For making felt

Uses of Rabbit Manure in Gardening

Can rabbit manure be used as fertilizer in the garden?

Rabbit manure is an exceptional form of manure. It has a rich, higher nutrient level, can be used fresh, and does not burn plant roots like other manure. It is just the right soil conditioner to use in any garden.

Here are some benefits of using it:

Nutrient-rich: Rabbit manure is twice as rich as chicken manure and contains four times higher the nutrient content of horse or cow manure.

Easy to work with: Rabbit manure doesn't have the same offensive odor as other manure. Being that it's in the form of little round pallets, you can handle it easily and apply it to your garden. It is also drier when compared to chicken manure.

Can be used fresh: You can apply rabbit manure directly to your garden without composting it first. Other manures, like chicken, cow, and horse dung, must be composted before being regarded as ready. If you use them fresh, they can burn your plant's roots. These manures need to be well-rotted, which takes up to three months.

Versatile: The rabbit manure pellets are used in ornamental flower beds and vegetable gardens. Furthermore, they are a rich nitrogen source to get a compost pile going and are used for up-dressing lawns.

No weed seed: Rabbit manure is usually obtained from pet rabbits not fed viable weed seed. This manure is taken from under the pet rabbit's hutches where the rabbits are kept. As sheep manure is so weedy, rabbit manure is weed-free when used in your garden. Ensure that your rabbit bedding material doesn't come close to the manure, which is why it is better to use bedding materials that are weed-seed-free.

Rabbit manure is affordable: The price tag is another wonderful benefit of using rabbit manure in gardening. You can get them locally or commercially through online retail outlets.

Rabbit manure is safe: Rabbit manure can be used around pets and plants in the house without endangering them with zoonotic diseases.

Rabbit manure is a 2-1-1 fertilizer: One main benefit of using rabbit manure is that it's a 2-1-1 fertilizer. It consists of Nitrogen, Potassium, and Phosphorus.

This makeup is just perfect for fostering the healthy growth of plants. Its plant benefits are seen from the sowing to the harvesting phase, as it supplies the needed nutrients for a resilient growth cycle.

These macronutrients are vital in rabbit manure:

- **Nitrogen:** This is needed for leafy green vegetative growth.
- **Phosphorus:** This is needed for fruiting, stem growth, and root formation.
- **Potassium:** This is needed for fruit ripening, flowering, and disease resistance.

Rabbit manure as soil conditioner: Rabbit manure is a good soil conditioner. As a source of organic matter, it enhances moisture and drainage retention and poor soil structure when buried in the soil. Due to its nutrient level, earthworms and microorganisms benefit from rabbit manure.

The Use of Rabbit Bones and Organs

The Use of Rabbit Organs

Rabbit organs are regarded as nature's nutrient-filled food. The reason is that there are many health benefits when you eat them. Rabbit liver is a reasonable size and has a mild to moderate flavor. Combined with other organs like the heart and kidney, they are tasty when fried with bacon and onions. You can grill the organs with onions and garlic, crush them into a paste, and spread them on crackers.

Here are the rabbit organs and their protein and vital nutrient content needed for your body.

- **Kidney:** It's rich in zinc, vitamins A, D, E, K, magnesium, iron, folate, and B vitamins, including B12.
- **Liver:** It's rich in potassium, zinc, vitamins A, B2, B6, B9, B12, D, C, E, calcium, magnesium, phosphorus, niacin, folate, choline, copper, and iron.
- **Heart:** It contains vitamins B6, B12, folate, iron, phosphorus, and copper. These are some of the benefits you get from the consumption of rabbit organs.

The Use of Rabbit Bones

Rabbit bones are rich in Potassium, Calcium, Magnesium, Phosphorus, and other minerals needed to develop and nourish your bones. Furthermore, rabbit bones enhance joint health.

Rabbit bone is a perfect flavor-based for broths, consommé, stock, and much more. This bone produces a silken and full-bodied broth, a good flavor enhancer for recipes. It is tasty when sipped on its own. You can boil your rabbit bones or roast them to increase their flavor. Rabbit stock, in any recipe, can take the place of water.

If you want that extra layer of savor when preparing potatoes, rice, lentils, or beans, consider including them in your dish and having them in your pantry as a major flavor enhancer.

The analogy of having your cake and eating it, in this case, is that you enjoy both the meat and its by-products because the benefits from all rabbit parts weigh more when compared to just the meat. So, next time you butcher a rabbit, know the skin, tail, head, droppings, etc., and are not just waste. They can yield more than you realize.

Bonus Chapter: Responsible Rabbit Raising: Ethics and Regulations

Congratulations! You've come a long way. You are now fully aware of what you need to know and have when raising rabbits for meat. However, there are still a few things to add - the ethics and regulations guiding the raising of rabbits.

Raising animals for food comes with a lot of responsibility. As a backyard farmer, it's up to you to ensure your rabbits live happily and healthily and are treated humanely. You'll need to make difficult decisions concerning how many litters to breed, how to dispatch the rabbits humanely, and how to sell or distribute the meat legally and ethically.

Raising rabbits for meat isn't a hobby; neither are rabbits just commodities to be sent off for food or income. It's a business that requires compassion and empathy.

The Moral Responsibility of Raising Animals for Food

Ethically, raising rabbits for meat is a big responsibility. By putting in the necessary effort to keep your animals healthy and happy and by striving to provide them with good, humane lives and deaths, you'll be able to enjoy the fruits of your labor with a clear conscience. There are some key things to keep in mind, which entail:

- **Doing Your Research**

Learn your rabbits' needs and make sure you can commit to meeting these needs. Proper housing, nutrition, handling, and healthcare are not optional. Therefore, research regulations regarding breeding, housing, and the regulations regarding selling meat in your area. The more you know, the better you can care for your rabbits.

Learn your rabbits' needs and make sure you can commit to meeting these needs.
https://www.pexels.com/photo/photo-of-a-woman-thinking-941555/

- **Focusing on Welfare**

Your rabbits should have good living conditions, opportunities for exercise, quality food, and veterinary care. Monitor them daily for signs of distress or disease and take action quickly. Handle them gently and move them calmly to avoid stress. Make sure any equipment used for their care is properly sized and maintained. Your rabbits' well-being must be the top priority.

- **Committing to Responsible Harvesting**

When it's time to butcher your rabbit, use the most humane methods you can, thus ensuring a quick and painless death. Have a plan in place and the proper tools for slaughtering. Remember, these animals have provided sustenance for you and your family, so they deserve your utmost respect, even at the end of their lives.

Providing Good Welfare and Humane Treatment

Providing a high standard of welfare and humane treatment for your animals is non-negotiable as a responsible rabbit farmer. Rabbits are living creatures that feel pain, fear, and distress, so they deserve your compassion.

Ensure your rabbits have spacious, well-ventilated housing that protects them from harsh weather. Provide opportunities for them to move about freely and plenty of mental stimulation. Feed them a healthy diet and give them constant access to fresh, clean water. Ensure your rabbits are monitored daily and taken to a vet as soon as needed.

Having raised these rabbits yourself, it's only right that you put them down as painlessly and quickly as possible. The most ethical option for butchering rabbits is cervical dislocation (breaking the neck), performed by a skilled operator. Some farmers prefer to hire a mobile slaughter unit to stun and kill the rabbits on-site. Whichever method you choose, make sure it causes a quick and painless death.

There are also regulations around breeding, housing, transporting, and selling rabbits and rabbit meat that you must follow. Research your city, county, and state laws to ensure you stay within legal limits for the number of does and litters permitted and requirements for selling meat. Some areas may require permits, licenses, or inspections.

As a responsible steward of animals and the environment, you must provide good welfare for the rabbits, use sustainable farming practices, and follow all regulations. By doing so, you'll feel proud of producing nutritious food in a kind and moral way.

Relevant Laws: Animal Welfare Act

Understanding legal regulations and the ethical responsibilities of raising rabbits should be a priority for backyard rabbit farmers. The Animal Welfare Act establishes rules for humane care and treatment of rabbits. Humanely treating rabbits isn't just about regulations. It's also about shaping a better world for your furry friends.

These regulations show how animals should be housed, handled, and given medical attention in laboratories or the vibrant arenas of circuses and zoos. Although the Animal Welfare Act prioritizes animals used for research, exhibition, and entertainment, it also casts a protective net over all creatures, even our beloved pets. This Act is a reminder that responsible ownership and thoughtful care extend to every corner of the animal kingdom.

The Animal Welfare Act isn't just a legal document; it's a commitment to compassion. It calls farmers to embrace empathy and kindness. As you familiarize yourself with its contents, you're committing to keeping up decent ethics and standards in the raising of rabbits.

Several states have additional laws for backyard rabbit farming. These laws cover breeding practices, selling meat, and animal cruelty. As a rabbit farmer, you have to understand these laws and regulations; carefully following them will help ensure you run an ethical, responsible operation. For example, some states prohibit the sale of uninspected meat, including rabbit meat. It's vital to check in with your local regulations.

Ethically, as a rabbit farmer, you must commit to responsible, humane farming practices that respect the rabbits' basic needs and natural behaviors. Some key principles include:

- **Providing Good Welfare**

Keep rabbits healthy, give them space to exercise, and practice positive reinforcement training.

- **Preventing Suffering**

Quickly treat injuries or illnesses, handle and transport rabbits carefully, and use humane slaughter methods.

- **Honoring Their Natural Living**

Give rabbits opportunities to socialize, forage, burrow, and play. Enrich their environment with tunnels, toys, and other stimuli.

- **Using Sustainable Practices**

Consider breeding for hardiness, mothering ability, and other useful traits. Avoid overbreeding.

By following these guidelines and maintaining high standards of care, you'll feel confident that you are acting with integrity as a backyard rabbit farmer.

Breeding Regulations

Most areas have regulations on breeding rabbits to prevent overpopulation and ensure good breeding practices. These may limit the number of litters a doe can have per year and prohibit inhumane caging conditions. Some states require breeders to be licensed and inspected.

Selling Meat Regulations

To sell rabbit meat, you must understand food production and sales regulations in your area. These typically cover:

- **Licensing and Inspections**

Most places require a license to sell meat and periodic inspections of your facilities.

- **Processing Requirements**

Meat must be processed in a licensed slaughterhouse or a government-approved personal facility. Also, follow specific rules that govern humane harvesting and handling.

- **Packaging and Labeling**

Meat must be properly packaged, labeled, and refrigerated or frozen to ensure safety and allow traceability. Labels provide information like weight, ingredients, producer details, and use-by date.

- **Zoning Laws**

These laws regulate where you can breed, raise, process, and sell rabbits in your community. Check with local authorities about requirements in your area.

Additional Considerations

Other regulations apply, such as transporting rabbits, importing new breeding stock, using pharmaceuticals, and disposing of waste. It's a good idea to check with organizations like the United States Department of Agriculture (USDA), the U.S. Food and Drug Administration (FDA), and the American Rabbit Breeders Association for the latest rules, regulations, and recommendations to follow.

Licensing Requirements for Commercial Rabbit Farms

There are certain licensing regulations you must follow as a commercial rabbit farmer. However, they vary in each country and region.

In the U.S., the United States Department of Agriculture (USDA) oversees regulations for commercial rabbit farms. Operations with over 3,000 rabbits must obtain a license, register with the USDA, and meet minimum standards of care under the Animal Welfare Act. Licensed farms are subject to unannounced inspections to check rabbits' health, housing, and humane handling.

Some of the key requirements for commercial farms include:
- Providing each rabbit with enough space to freely stand, lie, and turn around.
- Access to clean food and water daily.
- Proper heating, cooling, ventilation, and lighting systems.
- Regularly cleaning and disinfecting enclosures to keep rabbits healthy.
- Handling and harvesting rabbits according to American Meat Institute guidelines.

In addition to animal welfare regulations, there are strict requirements for selling rabbit meat for human consumption. They include:
- Meat must be processed in a licensed facility that follows proper sanitation and food safety procedures.
- Farms must keep detailed records to trace the origin and distribution of all meat sold.

As a responsible farmer, follow all regulations carefully and keep up-to-date with any changes in these regulations. Build positive relationships with inspectors and policymakers. Additionally, always remember that regulations exist to protect the welfare of your animals, the safety of consumers, and the sustainability of your farm. Following them is key to running an ethical operation.

Transporting and Handling Rabbits Legally and Ethically

Transporting and handling your rabbits in an ethical, responsible way is not only vital for their well-being but also a legal requirement. Your moral duty as a rabbit farmer is to provide humane care for your animals during all stages of their lives, including when you have to move or handle them.

When transporting your rabbits, you must follow these regulations:
- **Provide Food, Water, and Rest Periods**

In transit, rabbits must have access to food at least every 12 hours, water every six hours, and five-hour rest periods.

- **Use Proper Enclosures**

Transport cages must be constructed to protect rabbits from injury, contain waste, and allow them to stand up, lie down, and turn around. Wire flooring is prohibited.

- **Protection from Extreme Weather**

Transport vehicles must maintain temperatures between 45 to 85 degrees Fahrenheit. Rabbits must be shaded and have fans/sprinklers in warm weather.

- **Ensure Humane Handling**

Rabbits should be handled gently. Never hold them by their ears; use grasping areas over their back and rump instead. Dropping, kicking, or throwing rabbits is strictly prohibited.

When handling and moving your rabbits on the farm, be extremely careful. Rabbits can be easily stressed, frightened, and injured if not handled properly. Move them slowly and confidently, supporting their whole body. Never chase or make sudden movements and loud noises around them.

By following these regulations and using humane practices in your operations, you'll raise happy and healthy rabbits and build a sustainable business. Your customers will appreciate knowing their meat came from animals that were well cared for.

Slaughtering Rabbits Humanely: Methods and Regulations

Once your rabbits are ready for harvesting, it's vital to do so humanely and use methods that comply with regulations.

- **Cervical Dislocation**

This is the most common method of slaughtering rabbits. It involves quickly breaking the neck to sever the spinal cord, which kills the rabbit instantly. Proper training is required to perform this technique humanely and efficiently. However, it is prohibited in some areas, so check with your local regulations.

- **Electrical Stunning**

Electrical stunning is another option. This method uses a low-voltage current to stun the rabbit before bleeding it out. Specialized equipment is required, and there are strict guidelines on voltage, amperage, and length of stun. When done properly according to the law, this method is considered humane.

- **Bullet Dispatch**

For small farms, bullet dispatch is allowed and considered humane when performed by a skilled marksman with the proper firearm and ammunition. However, many areas prohibit discharging firearms and have additional regulations regarding storage, licensing, noise, and liability, which must be considered.

Guidelines for Humane Slaughter

- Rendering the animal unconscious and insensible to pain immediately.
- Do not restrain the animal in a way that causes injury or pain before unconsciousness.
- Check that the animal is unconscious and does not regain consciousness before death.
- Bleed the animal as soon as it is unconscious to ensure death.
- Provide proper training to anyone performing humane slaughter methods.
- Follow all local, state, and federal laws regarding humane slaughter and food safety.

Your rabbits deserve a quick and painless end, and your customers deserve safe, humanely produced meat. With diligence and compassion, you can achieve both.

Disposing of Rabbit Remains Properly and Legally

Disposing of your rabbits properly after slaughter is vital for legal and health reasons. As the farmer, you must handle remains correctly.

Proper Disposal Methods

The most common methods for disposing of rabbit remains are burial, incineration, and composting. Burying the remains at least two feet away from water sources is acceptable in many areas. However, some places have regulations against burying dead livestock, so check local ordinances.

Incinerating the remains in an approved incinerator is also an option. However, the equipment for this method can be expensive, and permits may be required. Composting the remains in a secure composting bin with carbon-rich brown materials like sawdust, straw, and leaves is a sustainable method, but it can take 6-12 months for remains to break down fully. The compost should not be used on food crops.

Regulations and Zoning Laws

Most areas prohibit dumping rabbit remains in landfills, waterways, and open areas. There are strict laws regarding the disposal of dead livestock to prevent the spread of diseases and pests. It's important to understand the regulations in your city or county to avoid hefty fines or legal trouble. Zoning laws may also prohibit composting and incinerating remains in residential areas. Always check with local authorities regarding which disposal methods are allowed and any permits needed on your property.

An Ethical Obligation

As a rabbit farmer, treat the remains as you would want your remains to be handled. Keeping records of disposal, composting, and incineration at high enough temperatures and properly securing remains from pests and predators are all responsible and ethical disposal practices. How you handle remains says a lot about your standard of care and respect for the animals in your charge. Do right by your rabbits even after they are gone.

There you have it - the ins and outs of responsible and regulated rabbit farming. You are responsible for your animals' welfare and must operate legally and ethically as a rabbit farmer. Provide good care, humane handling, and a sustainable living environment. Educate yourself on regulations and reflect carefully on the ethics of raising animals for food.

With diligence and compassion, you can raise rabbits responsibly without losing sight of your moral duty toward them.
https://www.pexels.com/photo/smiling-girl-holding-gray-rabbit-1462636/

With diligence and compassion, you can raise rabbits responsibly without losing sight of your moral duty toward them. Raising animals is a big responsibility, but if done correctly, it can be a rewarding experience for you and your community. The more you understand ethical farming practices, the better-equipped you are to make good decisions and set a positive example.

Conclusion

As you wrap up your learning of raising rabbits for meat, you must address a sentiment many people have grappled with - the undeniable cuteness of these furry creatures. It's completely normal to feel a twinge of hesitation when it comes to harvesting animals you've cared for. However, as you've seen throughout this guide, practical considerations and valuable insights can help you find a balance between your emotions and your goals.

Like any aspect of homesteading, Rabbit rearing comes with its own challenges and rewards. Choosing the right breed for your specific needs is the foundation of success. Be it for meat production or specific traits, understanding breed characteristics is key. Providing appropriate housing, nutrition, and medical care goes a long way in ensuring your rabbits lead healthy and productive lives. Remember that raising rabbits isn't just about physical needs - it's also about emotional well-being. Spending quality time with your rabbits, observing their behaviors, and creating a stress-free environment can contribute significantly to their overall health and contentment.

When the time comes for processing, it's essential to approach it with respect and compassion. Employing humane harvesting techniques and utilizing as much of the rabbit as possible demonstrates a commitment to ethical practices. While the path of rabbit farming for meat is undoubtedly rewarding, it's crucial to tread carefully and be aware of potential pitfalls. Maintaining a rigorous disease prevention regimen is non-negotiable. Rabbits are susceptible to various illnesses, so staying informed about potential health risks and implementing preventive measures can save you from heartache down the road.

Breeding should always be approached with a clear purpose and a commitment to improving the breed. Overbreeding or neglecting to consider genetic factors can lead to unintended health issues in future generations. Emotionally, preparing yourself for the harvest is an aspect that can't be overlooked. It's okay to have mixed feelings, but acknowledging and reconciling these emotions is essential to maintaining a healthy perspective.

Approach each step of this journey as an opportunity to learn and grow. There will be successes and challenges, each contributing to your experience and expertise. Try to create a balance between your emotional connection to the rabbits and the practical purpose they serve. Remember that raising animals for meat is a responsible choice that contributes to sustainability and self-sufficiency.

Here's another book by Dion Rosser that you might like

References

(2008, March 20). Ethical breeding: 10 golden rules. Tru-Luv Rabbitry: Quality Holland Lops in Malaysia. https://truluvrabbitry.com/2008/03/20/ethical-breeding-10-golden-rules/

(2021, February 1). Slaughter: how animals are killed. Viva! The Vegan Charity; Viva! https://viva.org.uk/animals/slaughter-how-animals-are-killed/

(N.d.). Fao.org. https://www.fao.org/3/t1690e/t1690e.pdf

(n.d.). Killing rabbits for food. 3 best ways to kill a rabbit. Raising-rabbits.com. https://www.raising-rabbits.com/killing-rabbits.html

(n.d.). Recommended methods of euthanasia: Rabbits. Umaryland.edu. https://www.umaryland.edu/media/umb/oaa/oac/oawa/guidelines/Euthanasia_Rabbits_12.2020.pdf

(n.d.). Slaughtering and dressing rabbits. Msstate.edu. http://extension.msstate.edu/content/slaughtering-and-dressing-rabbits

(N.d.). Standardmedia.Co.Ke. https://www.standardmedia.co.ke/farmkenya/amp/article/2001340660/how-to-make-better-use-of-rabbits-by-products%2010

10 of the most common rabbit health emergencies. (2020, April 16). Best4Bunny. https://www.best4bunny.com/10-of-the-most-common-rabbit-health-emergencies/

Alyssa. (2019, November 4). What do you get from a meat rabbit? Homestead Rabbits. https://homesteadrabbits.com/meat-rabbit-parts/

Alyssa. (2019, October 11). Commercial meat rabbit growth rates. Homestead Rabbits. https://homesteadrabbits.com/meat-rabbit-growth-rates/

Alyssa. (2022, March 4). Raise Meat Rabbits: Quick start guide. Homestead Rabbits. https://homesteadrabbits.com/raise-meat-rabbits/

Alyssa. (2022, March 4). Raise Meat Rabbits: Quick start guide. Homestead Rabbits. https://homesteadrabbits.com/raise-meat-rabbits/

Baby rabbit information. (n.d.). Com.au. https://mtmarthavet.com.au/baby-rabbit-information/

Barnett, T. (2020, April 13). Can you keep rabbits outdoors: Tips for raising backyard rabbits. Gardening Know How. https://www.gardeningknowhow.com/garden-how-to/beneficial/can-you-keep-rabbits-outdoors.htm

Brad. (2021, January 3). Flemish Giant Rabbits: Care and Breeding. Northern Nester. https://northernnester.com/flemish-giant-rabbits/

Browning, H., & Veit, W. (2020). Is humane slaughter possible? Animals: An Open Access Journal from MDPI, 10(5), 799. https://doi.org/10.3390/ani10050799

Budnick, T. (n.d.). A "hare" raising lapse in meat industry regulation: How regulatory reform will pull the meat rabbit out from welfare neglect. Animallaw.Info. https://www.animallaw.info/sites/default/files/Rabbit%20Meat%20%26%20Regulatory%20Reform.pdf

Californian rabbit characteristics, origin, uses. (2022, January 31). ROYS FARM. https://www.roysfarm.com/californian-rabbit/

Caring for an older rabbit. (n.d.). Org.uk. https://www.rspca.org.uk/adviceandwelfare/pets/rabbits/senior

Carter, L. (2020, May 3). How to take care of baby bunnies. Rabbit Care Tips; Lou Carter. https://www.rabbitcaretips.com/how-to-take-care-of-baby-bunnies/

Code of practice for the intensive husbandry of rabbits. (2020, June 23). Agriculture Victoria. https://agriculture.vic.gov.au/livestock-and-animals/animal-welfare-victoria/pocta-act-1986/victorian-codes-of-practice-for-animal-welfare/code-of-practice-for-the-intensive-husbandry-of-rabbits

Collaborator, B. (2020, January 30). What temperature is too cold for rabbits? K&H Pet Products. https://khpet.com/blogs/small-animals/what-temperature-is-too-cold-for-rabbits

Creating a good home for rabbits. (n.d.). Org.uk. https://www.rspca.org.uk/adviceandwelfare/pets/rabbits/environment

Creating the ideal home for your rabbits. (n.d.-b). Org.uk. https://www.pdsa.org.uk/pet-help-and-advice/looking-after-your-pet/rabbits/creating-the-ideal-home-for-your-rabbits

Crossbreeding, outcrossing, linebreeding, and inbreeding. (n.d.). LOTS OF LOPS RABBITRY. http://www.lotsoflops.com/crossbreeding-outcrossing-linebreeding-and-inbreeding.html

Dec. (2017, December 21). Five common diseases that affect rabbits. Petmd.com; PetMD. https://www.petmd.com/rabbit/conditions/five-common-diseases-affect-rabbits

Diina, N. (n.d.). Farm4Trade Suite. Farm4tradesuite.com. https://www.farm4tradesuite.com/blog/10-reasons-to-start-raising-rabbits

Flemish Giant rabbit: Characteristics, uses, origin. (2022, January 31). ROYS FARM. https://www.roysfarm.com/flemish-giant-rabbit/

Guidelines for keeping pet rabbits. (2020, June 12). Agriculture Victoria. https://agriculture.vic.gov.au/livestock-and-animals/animal-welfare-victoria/other-pets/rabbits/guidelines-for-keeping-pet-rabbits

How to skin a rabbit: 2 easy methods (with pictures). (2009, May 2). WikiHow. https://www.wikihow.com/Skin-a-Rabbit

Humane slaughter: how we reduce animal suffering. (2014, May 27). World Animal Protection. https://www.worldanimalprotection.org/historical-achievements/humane-slaughter

Is rabbit manure good to use in the garden? (2020, July 15). Deep Green Permaculture. https://deepgreenpermaculture.com/2020/07/15/is-rabbit-manure-good-to-use-in-the-garden/?amp=1

Jollity. (2020, February 7). Rabbit lifespan and life stages. Oxbow Animal Health. https://oxbowanimalhealth.com/blog/rabbit-life-stages/

Jones, O. (2020, April 15). 10 best meat rabbit breeds in the world (2023 update). Pet Keen. https://petkeen.com/best-meat-rabbit-breeds/

Kathryn. (2013, December 23). Colony raising rabbits: How to get started. Farming My Backyard. https://farmingmybackyard.com/colonyraisingrabbits101/

Kathryn. (2013, December 23). Colony raising rabbits: How to get started. Farming My Backyard. https://farmingmybackyard.com/colonyraisingrabbits101/

Kathryn. (2019, December 12). Which meat rabbit breeds should you raise? Farming My Backyard. https://farmingmybackyard.com/meat-rabbit-breeds/

Kathryn. (2019, May 29). The best ways to feed rabbits (besides pellets)! Farming My Backyard. https://farmingmybackyard.com/feed-rabbits/

Kellogg, K. (2022, January 4). How to tan a rabbit hide. Mother Earth News – The Original Guide To Living Wisely; Mother Earth News. https://www.motherearthnews.com/diy/how-to-tan-a-rabbit-hide-zmaz83jfzraw/

Klopp, J. (n.d.). Bunny Farming: Why Do People Farm Rabbits? Is It Cruel? Thehumaneleague.org. https://thehumaneleague.org/article/bunny-farming

Kruesi, G. (2020, January 3). Staying warm with rabbit wool. Chelsea Green Publishing. https://www.chelseagreen.com/2020/staying-warm-with-rabbit-wool/

Martin, A. (n.d.). Caring for the elderly or senior rabbit. Lafeber.com. https://lafeber.com/mammals/caring-for-the-elderly-or-senior-rabbit/

McClure, D. (n.d.). Disorders and Diseases of Rabbits. MSD Veterinary Manual. https://www.msdvetmanual.com/all-other-pets/rabbits/disorders-and-diseases-of-rabbits

Montano, C. (2021, January 18). Bone Broth or rabbit. Christinamontano.com. https://www.christinamontano.com/amp/bone-broth-or-rabbit

Murphree, M. E. (n.d.). Backyard grower-consumer perceptions of rabbit meat consumption in rural Mississippi al Mississippi. Msstate.edu. https://scholarsjunction.msstate.edu/cgi/viewcontent.cgi?article=6542&context=td

Ned, & Hannah. (2023, February 13). 6 surprising rabbit manure benefits. The Making Life. https://themakinglife.com/rabbit-manure-benefits/

New Zealand rabbit characteristics use origin. (2022, January 31). ROYS FARM. https://www.roysfarm.com/new-zealand-rabbit/

NOSE TO TAIL-uses for every part of the domestic rabbit. (2012, February 11). Rise and Shine Rabbitry. https://riseandshinerabbitry.com/2012/02/11/nose-to-tail-uses-for-every-part-of-the-domestic-rabbit/

Ockert, K. (2015, November 10). MSU extension. MSU Extension. https://www.canr.msu.edu/news/determining_cage_size_for_rabbits

Owuor, S. A., Mamati, E. G., & Kasili, R. W. (2019). Origin, genetic diversity, and population structure of rabbits (Oryctolagus cuniculus) in Kenya. BioMed Research International, 2019, 7056940. https://doi.org/10.1155/2019/7056940

Pellets and nutrition for meat rabbits. (2012, May 23). Rise and Shine Rabbitry. https://riseandshinerabbitry.com/2012/05/23/pellets-and-nutrition-for-meat-rabbits/

Peoria zoo. (2014, April 7). Peoria Zoo. https://www.peoriazoo.org/animal-groups/mammals/giant-flemish-rabbit/

Planning a Homemade Rabbit Cage. (2014). Therabbithouse.com. http://www.therabbithouse.com/indoor/designing-rabbit-cage.asp

Poindexter, J. (2017, February 23). How to butcher a rabbit humanely in 6 quick and easy steps. Morning Chores. https://morningchores.com/how-to-butcher-a-rabbit/

Pratt, A. (2019, November 11). 5 life stages of pet rabbits and how to keep them healthy. The Bunny Lady; Amy Pratt. https://bunnylady.com/rabbit-life-stages/

Pratt, A. (2020, March 6). How to make Critical Care rabbit formula for emergencies. The Bunny Lady; Amy Pratt. https://bunnylady.com/critical-care/

Pratt, A. (2021, April 5). Rabbits need more space than you think. The Bunny

Lady; Amy Pratt. https://bunnylady.com/space-for-rabbits/

Pratt, A. (2021, March 8). How big do rabbits get? (smallest and largest breeds). The Bunny Lady; Amy Pratt. https://bunnylady.com/how-big-do-rabbits-get/

Preparing for emergencies. (n.d.). Therabbithaven.org. https://therabbithaven.org/preparing-for-emergencies

Rabbit bones. (n.d.). Steaksandgame.com. https://www.steaksandgame.com/rabbit-bones-1458

Rabbit breeding system. (2020, April 12). McGreen Acres. https://mcgreenacres.com/blog/rabbits/rabbit-breeding-system

Rabbit breeds: Best 17 for highest profits. (2022, January 28). ROYS FARM. https://www.roysfarm.com/rabbit-breeds/

Rabbit farming: Best beginner's guide with 28 tips. (2022, January 7). ROYS FARM. https://www.roysfarm.com/rabbit-farming/

Rabbit personalities and lifespan. (n.d.). The Anti-Cruelty Society. https://anticruelty.org/pet-library/rabbit-personalities-and-lifespan

Rabbit stock. (2008, January 28). Saveur. https://www.saveur.com/article/Recipes/Rabbit-Stock/

Rabbit's life cycle: From bunny to adult. (n.d.). CYHY. https://creatureyearstohumanyears.com/resources/rabbit-life-cycle

Raising meat rabbits. (2016, October 14). Farming My Backyard. https://farmingmybackyard.com/rabbits/

Richardson, H. (2022, June 8). How to know when to cull rabbits. Everbreed. https://everbreed.com/blog/how-to-know-when-to-cull-rabbits/

Shy rabbits. (2011, July 10). House Rabbit Society. https://rabbit.org/2011/07/faq-shy-rabbits/

Składanowska-Baryza, J., Ludwiczak, A., Pruszyńska-Oszmałek, E., Kołodziejski, P., & Stanisz, M. (2020). Effect of two different stunning methods on the quality traits of rabbit meat. Animals: An Open Access Journal from MDPI, 10(4), 700. https://doi.org/10.3390/ani10040700 (72), K. (2018, September 18). How to skin a rabbit – A step-by-step guide. Steemit. https://steemit.com/howto/@ketcom/how-to-skin-a-rabbit-a-step-by-step-guide

Suitable environment for rabbits. (2015, November 20). Nidirect. https://www.nidirect.gov.uk/articles/suitable-environment-rabbits

Sullivan, K. (2019, November 26). Is your rabbit sick? 9 signs the answer may be "yes." PETA. https://www.peta.org/living/animal-companions/is-my-rabbit-sick/

Tertitsa, T. (2013, October 27). Rabbit stewardship: Ethical, humane, conscientious raising/husbandry. One Community Global. https://www.onecommunityglobal.org/rabbits/

The Backyard Rabbitry. (2023, February 6). How to choose the right rabbit breed for meat production. The Backyard Rabbitry. https://thebackyardrabbitry.com/how-to/how-to-choose-the-right-rabbit-breed-for-meat-production.html

Vanderzanden, E., & Kerr, S. (n.d.). Raising rabbits for meat: Providing basic care. Oregonstate.edu. https://catalog.extension.oregonstate.edu/sites/catalog/files/project/pdf/ec1655.pdf

Vanderzanden, E., & Kerr, S. (n.d.). Raising rabbits for meat: Providing basic care. Oregonstate.edu. https://catalog.extension.oregonstate.edu/sites/catalog/files/project/pdf/ec1655.pdf

Walker, J. (2015, January 29). Keeping pregnant rabbits healthy, safe and warm. Coops and Cages. https://www.coopsandcages.com.au/blog/keep-pregnant-rabbits-safe-healthy-warm/

What to feed meat rabbits. (2019, February 14). Countryside. https://www.iamcountryside.com/homesteading/feed-meat-rabbits/

What to know about New Zealand rabbits. (n.d.). WebMD. https://www.webmd.com/pets/what-to-know-about-new-zealand-rabbits

What to know about the Californian rabbit. (n.d.). WebMD. https://www.webmd.com/pets/what-to-know-about-californian-rabbits

What to know about the Flemish giant rabbit. (n.d.). WebMD. https://www.webmd.com/pets/flemish-giant-rabbit

Printed in Dunstable, United Kingdom